D0421412

War and the city

War and the city

G. J. Ashworth

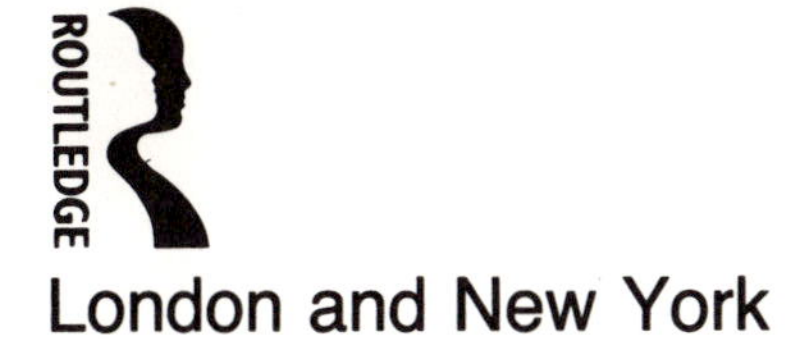

London and New York

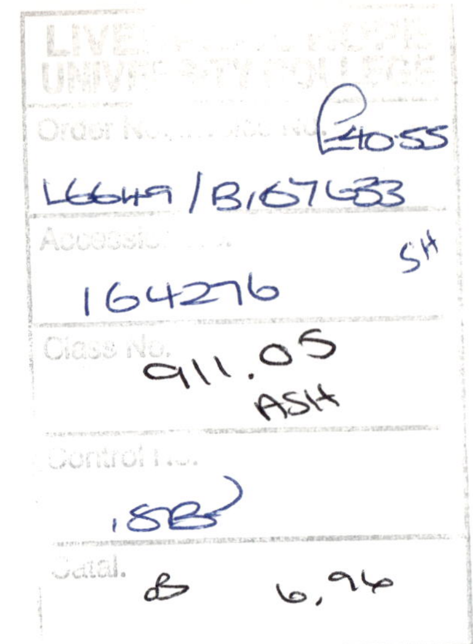

First published 1991
by Routledge
1 New Fetter Lane, London EC4P 4EE

Simultaneously published in the USA and Canada
by Routledge
a division of Routledge, Chapman and Hall, Inc.
29 West 35th Street, New York, NY 10001

Filmset by J&L Composition Ltd, Filey, North Yorkshire
Printed in Great Britain by Mackays of Chatham PLC,
Chatham, Kent

British Library Cataloguing in Publication Data
Ashworth, G. J. (Gregory John) *1941–*
War and the city.
1. Cities. Effects of war
I. Title
355.020091732

ISBN 0–415–05347–1

Library of Congress Cataloging in Publication Data
Ashworth, G. J. (Gregory John)
War and the city / G. J. Ashworth.
p. cm.
Includes bibliographical references and index.
ISBN 0–415–05347–1
1. Cities and towns–History. 2. War–History. 3. Military geography. I. Title.
HT113.A84 1991
355.4'71732—dc20 90–26361
CIP

Contents

Figures

Tables

Preface

I am conscious of my good fortune – an accident of time and place of birth – in never having been called upon to fight in wars, or even be trained to do so, never having fired any weapon outside a fairground or been fired upon, and never having worn the uniform of any armed service or security force. This is a declaration of fact, not faith, which has resulted in my experience with the subject of this book being completely at second hand.

I have therefore been particularly dependent upon the experiences of others. My main sources of information and inspiration have been the writings and personal communications of practitioners and commentators from many academic disciplines, not all of whom are adequately acknowledged in the text. To these must be added two other less conventional sources, namely visits to the locations of many of the conflicts described here and to the remaining artefacts of defence in many parts of the world and simulational gaming.

Among the many who have fed me useful titbits of information, ransacked their personal libraries or offered their comments upon parts of the text, the following people must be mentioned: John Bradbeer, Michael Damminga, Ray Riley, John Tunbridge, Chris van Welsenes and Harry Vianen. Eric Runau and the cartographic section of the R.U. Groningen have drawn most of the maps with their usual skill. Finally, I must acknowledge the extraordinary patience of many members of my family, who have been trailed around so many battlefields and bunkers – in particular my son, Luke, whose work as a political scientist is acknowledged in the references, but whose most valuable contribution has been to refight all these battles with me.

There is a particular difficulty in writing about war (which is not overcome by adding its converse, 'peace'), and the serious charge can be laid that by writing about war it may be at best rendered acceptable and at worst encouraged. Should not the accent be shifted to peace and the title read, 'Defence and the city: how to render it unnecessary'?

One solution is to retreat behind technical terminology and an ostensibly scientific approach. This dehumanizes conflict and portrays it as an abstract game played to mutually acceptable rules. The alternative, the 'horrors of war' approach, which dwells upon the suffering and destruction inherent in war, may be voyeuristic, desensitizing, and thus ultimately counter-productive.

The charge that potential practitioners of urban conflict will find this a 'how-to-do-it' handbook is easily dismissed. It could be said that readers might include: nascent urban guerrillas seeking to glean instructions about where to build a barricade; urban rioters hoping to be forewarned about police crowd control tactics; or military commanders seeking some tips useful in a career of urban conquest. However, all such incidental detail is drawn from freely available sources – which would prove much more useful for such purposes.

Two defences are offered to the charge of encouragement. First, that of Terence: 'Nothing human can be strange to me'. Organized violence has been and continues to be an activity of people in cities. The study of war no more encourages it than studies of urban crime or prostitution are supportive of those activities. Such aspects of a flawed humanity cannot be ignored because they are unpleasant. There may even be a less neutral defence on the grounds that war will never be prevented by those who know nothing of it. Second, the relating of something of the reality of urban warfare, annihilation, and terrorism could conceivably make their repetition less likely, and to this hope, however naïve, this book must be dedicated.

Chapter 1

Introduction

The theme of this book is ostensibly simple: it examines a set of relationships between an activity (defence) and a particular spatial setting (cities).

Such relationships may be viewed from the standpoint of the defence function, in which case the title is best expressed as 'the urban factor in defence'. In such a study the city is seen as one environmental setting, among many others, whose distinguishing characteristics are variables influencing the way such an activity is pursued ('the urban geographical factor in defence studies') and are themselves liable to be influenced by its practice ('the impact of defence upon the city'). Conversely, the relationship may be examined from the standpoint of the city, in which case defence is examined as one of the many activities exercised in cities and the focus is upon the place of this function within the urban synthesis. The result would be 'an urban defence geography', in the same sense as existing urban social, economic, or political geographies. The chosen title, 'war and the city', indicates that both viewpoints will be taken, and on occasion it is in practice difficult to separate the two, however logical such a distinction may be. Nevertheless, the emphasis must remain firmly on the relationship between the two.

The justification for this approach rests upon two assertions.

1 A focus upon this link between the defence function and the urban setting reveals aspects of both that would not be apparent in studies focusing exclusively on one or the other. In other words, urban studies without a consideration of defence and defence studies without a consideration of cities are both seriously incomplete.
2 Such an examination has not previously been satisfactorily attempted.

The book as a whole must stand as surety for the first assertion, failing in its primary purpose if insufficient evidence is marshalled to be convincing on this point. At this stage a prima-facie case is simply stated: an activity which has occupied such a large part of the attention and resources of cities *and* has been an important factor in both their origins and growth (or, conversely, their decline and even annihilation) can be assumed to be worthy of at least

as much attention as has been devoted to other urban functions. The second assertion needs closer examination here.

MILITARY SCIENCE AND URBAN DEFENCE

There is no shortage of specialist literature upon either cities or defence. On the contrary, the study of various aspects of cities has for some time been a major focus of interest for many academic disciplines, for practitioners in urban planning and management, and for informed citizens and visitors. Similarly, defence has generated an enormous interest from historians, strategic and tactical studies specialists, technical specialists, and has attracted the interest of a much wider public than those professionally involved in it. 'The popularity of military history is such that it has ... assumed the proportions of a minor industry' (Holmes 1985). To which can be added popular interest in defence technology, military gaming and simulation, and military tourism, so that the only disagreement may be over the use of the adjective 'minor'.

> War has had us in its thrall. It has horrified us and fascinated us... The stench of war has seeped into our souls. We have talked endlessly about peace: but in the recesses of our imagination we have brooded often feverishly on war, and we have written about it more copiously than any previous generation.
>
> (Connell 1965: 221)

It would be strange indeed if these two prolific streams of academic, professional and popular literature did not touch, from time to time, upon the interrelationships of cities and defence. This being the case, then why has neither an 'urban geography of defence' nor an 'urban defence geography' emerged? This charge of bilateral neglect needs modification, explanation and mitigation.

Geography in military science

The acknowledged founding father of military science, Carl von Clausewitz, paid some attention to space, both as ground to be manoeuvred over and as an environment with characteristics that may influence the conduct of military operations (Greene 1943). But the nature of the conflict location was seen as a minor contributory variable compared with the major determining factors such as leadership, morale and strategic soundness. Subsequent commentators largely followed this lead, which in Germany resulted in a curious ambivalence. On the one hand, a 'scientific' general staff was developed, many of whose members were either formally trained in universities as geographers or were strongly influenced by the geopolitical ideas developed by university geographers, of which the most renowned was

Friedrich Ratzel. On the other hand, there was little recognition of the role of geography as such in military science. Penck in 1916 summed up the value of academic geography to the war effort then in progress, in a paper to the Berlin Academy. He produced a long list of contributions which on closer examination consist of either practical skills, such as military cartography, or vague notions of strengthening feelings of national unity. There is no mention of the role of geography in either strategy or tactics. This is despite almost a century's development of German war plans in which essentially geographical solutions were sought for the perceived geopolitical disadvantages of a land-locked and potentially surrounded German empire. The plans of Moltke senior, that led to victory in 1871, and von Schlieffen, whose plan in a modified form dictated German two-front strategy in 1914, can be traced back to the ideas of academic geographers, especially Ratzel (Wanklyn 1961).

Such influences were also largely confined to Germany, where both geography as a university discipline and a scientific approach to military staff work were uniquely developed and there were few imitators. In Britain Maguire's (1899) *Outlines of Military Geography* was mainly a chronological account of various field campaigns and a history of fortification which made little attempt to contribute to either strategy or tactics. Even in The Netherlands, where it might be expected that German influence would be strong, textbooks on 'military geography' (*Militaire Aardrijkskunde*) such as that of Wilson (1934) are, in practice, handbooks of the location and design of defence works for the use of engineers. In most countries geography was seen at most as contributing some specific skills of value to the technical specialists – in military cartography, engineering and artillery.

The attitudes of military science towards cities in particular can again be traced back to Clausewitz. He regarded the city as a battlefield variant, one among many in a checklist of possible environments, within which the theory and practice of war would be exercised. Little attention is paid to the characteristics of this environment and the implication of this neglect is that it is unsuitable and to be avoided. Subsequent military writers did little to alter this view. Two explanations can be offered for this.

First, until quite recently the urbanized portion of the world, and even of western Europe, was a small portion of the total. Peltier and Pearcy justify the neglect of the city by military science with an historical argument:

> As long as urban areas were small, or could be easily bypassed, there was little reason to consider an urban kind of military theater. Under the changing importance of urban areas and the growing extent of cities and suburbs, however, it does seem useful not only to distinguish this type of environment but to characterise its specific effect on military strategy.
>
> (Peltier and Pearcy 1966: 119)

Leaving aside the very tentative and reluctant acceptance that contemporary conflict may occur in cities, it can be objected that cities in the past, far from being bypassed, were frequently fought over, and additional arguments must be sought. Second, an aversion to fighting in towns is inherent in professional armies for various reasons (which will be argued at length later). In particular, the practice of military science as developed in the post-Napoleonic period required close command control so that the principles of strategic and tactical manoeuvres could be applied. Such leadership is particularly difficult to exercise in urban environments, which also possess the complications of the presence of civilians. Cities were avoided as both poor test beds for the ideas of the theorists and even as rather improper places for battles to occur.

Although fighting in the city was thus largely ignored by the military theorists, the practical defence of cities was a primary task of the military engineers, and the science (including the geography) of urban fortification remained important, especially in continental Europe, until well into the twentieth century.

Military science in geography

The charge of neglect of defence within geography can only be partially sustained. In the same way that the nineteenth-century German general staff in their adoption of a scientific approach to strategy had recruited the services of geography, there was also a detectable tendency for British geographers of the period to develop similar interests for similar reasons. If the physical position of the new German empire, especially in relation to its potential antagonists, encouraged the development of a strategy of manoeuvre on a continental scale, then the growing British consciousness of the possession of a global empire during the nineteenth century was a stimulus to strategic thinking on a world-wide scale. The main difference was that in Germany geography was regarded as containing skills useful for future staff officers and was therefore incorporated into the military curriculum. In Britain it was campaigning individuals who pleaded for the relevance of geography to the defence needs of the state. Indeed, the need to defend a world-wide Empire, and its supporting trading routes, was used as an argument for the establishment of geography as a university teaching discipline, and the example of Germany was used as a warning of British backwardness in this respect.

The most influential of these advocates was Mackinder, whose famous textbook, *Britain and the British Seas* (1902) has this approach implicit in its title. The theme of defence runs strongly through the book; geography having two important branches, economic geography and military geography, or the production of wealth and its defence from covetous neighbours. Military geography in turn has two aspects: 'strategic geography' (the conduct of campaigns), and 'tactical geography' (the conduct of battles). The latter is disregarded as of interest to the professional military, but the former is the basis of 'statesmanship' (Mackinder 1902: 314). In fact, much of Mackinder's writing in political and empire matters was a thinly disguised plea to statesmen for the application of the geographer's global view to the strategic defence of British world interests, an approach which itself owed much to Mahan's (1890) studies of the exercise of sea power and to Ratzel's (1903) broad-brush views of territory and power.

In more recent years there have been sporadic attempts to revive a 'military geography', most usually by political geographers developing the 'strategic' line of Mackinder. Cole's *Imperial Military Geography*, first published in 1924 but going through eight editions in the next ten years, was subtitled 'the general characteristics of the empire', by which he meant world geography, 'in relation to its defence'. Less specific in its purpose was Peltier and Pearcy's (1966) *Military Geography*, which was principally a geographical description of those zones in the world where political tensions had been reflected in military conflict. Mellor (1987) refined and developed this idea as the application of military considerations to the spatial organization of territory, which he has illustrated in a wide range of geographical and historical circumstances. Others, such as Cohen (1971) or O'Loughlin and van de Wusten (1986), have developed such a 'shatter-belt view' of the world – the task of geographers being to seek out and describe those regions which, 'because of complex national, religious, political, ideological, economic or physical components are areas of tension' (O'Loughlin and van de Wusten 1986: 493). Little more than passing reference is paid, however, to Mackinder's 'tactical geography' (the effects of environments upon the conduct of conflict), although Peltier and Pearcy (1966: 118) admit the necessity 'not only to distinguish the type of environment but to characterise its specific effect on military strategy'.

It was left, however, to O'Sullivan and Miller's (1983) *Geography of Warfare* and Faringdon's (1989), *Strategic Geography* to combine a world-scale strategic geography with investigations of the effects of the environmental variable on the conduct of both regular and guerrilla operations. O'Sullivan and Miller attempt to sketch the dimensions of a geography of war, while Faringdon relates political factors and weapon technology in actual and potential 'trouble spots'. Both books contrast sharply with the approach of Bateman and Riley's (1987) *Geography of Defence*, which is a series of essays dominantly on the detailed impacts of various defence functions upon local cases, the majority of which are both British and urban.

There have, however, been two multidisciplinary lines of research in the social sciences, that have developed over the last 20 years, to which geographers have contributed, and which have in turn contributed to a geography of defence. The first is the interest, principally among economists, in the military-industrial complex – i.e. the growth of a substantial economic sector, principally in the United States, as a result of government-initiated defence procurement spending. The effects of such concentrations of spending upon local economies has a demonstrable significance at the regional and urban level, resulting in what Letchin (1984) called the 'martial metropolis'. However, such studies generally have more in common – in terms of locational choices and methods of assessing local economic impacts – with other aspects of the economy than with the defence function as such.

The second theme, which is also a response to the political and military developments of the post-war decades, is usually termed 'peace studies'. This involves more than the addition of the word 'peace' to studies of its converse, war; it is generally based upon a strong implicit criticism of the assumptions and practice of two-power nuclear confrontation, and contains an advocacy of non-violent solutions to political conflict. It represents a scientific and educational response to political movements for disarmament, especially nuclear disarmament, and has been established as a university discipline in a number of institutes in north-western Europe. The origin and nature of this approach has principally attracted specialists in political science, international relations, and international law but a handful of specifically geographical studies have also emerged, notably the various publications of Openshaw (1983; 1985) and those edited by Pepper and Jenkins (1985). The geographical contribution has been particularly notable in the studies of the effects of future nuclear war, not least upon cities, and the monitoring of the efficacy of official policies for civilian protection in this event.

Most of the studies mentioned above, whether written from a geographical perspective or not, contain much material that is specifically of relevance to cities. They do not, however, amount to an urban defence geography.

There have been many attempts to produce classifications of cities based on the functions they fulfil and the consequent implications of these for site selection and morphological development. Indeed, the creation of such taxonomies was a central obsession of urban geographers for the first few generations of the study of the topic (Aurousseau 1924, Taylor 1949). Defence has frequently been included in the lists of functions to be considered, either in terms of a particular activity of a few specialized towns that could not comfortably be included under some other category, or as a factor to be arrayed alongside others – whether economic, social, administrative, religious or recreational – in explaining the origin, and shaping the development, of urban form and function. One of the earliest and most complete accounts is found in Mackinder (1902), where a classification of towns in Britain with a defence function is not only exhaustive in its categories but is coupled with explanations of location. Thus, the location of naval bases, dockyards, harbours of refuge, academies and depots are related to naval strategy and weaponry in the era before dreadnoughts, mines and submarines. Similarly, the location of a wide range of army functions is explained in terms of sets of locational variables that a later generation of geographers would call a model of the national distribution of defence towns.

Subsequently, however, no attempt was made to update or develop these ideas, which have remained dormant for two generations. Even as late as 1966, Peltier and Pearcy's *Military Geography*, which includes a short section entitled 'Cities as military regions' (with the justification that cities are 'a special condition among global environments'), does little more than paraphrase Clausewitz 150 years later.

> The (urban) terrain is something like a region of strongly compartmentalised badland topography honeycombed with caves, but with one difference: artillery and aerial bombardment made the terrain more difficult to negotiate by building up piles of rubble.
>
> (Peltier and Pearcey 1966: 112)

Although defence has not been ignored then, it has received less attention than most other urban functions, including many urban activities and phenomenon that are both more transient and intrinsically more trivial than defence. An explanation of this may lie in the philosophy and methodology of the social sciences as they have developed their investigations in cities over the last 20 years.

Despite the overall demonstrable importance of defence, its impact upon cities tends to be sporadic. War may have a more dramatic impact on the city than, for example, the purchase of shoes or the visits of tourists, but it occurs less often, is less measurable in its incidence, is less predictable and is generally beyond the influence of urban planners and managers. Therefore, given the methods of investigation that have prevailed in most urban studies for a generation, it is not surprising that more has been written on the locational patterns, impacts and planning of shoe shops or tourism facilities than on the effects of war or riot.

DEFENCE AND THE CITY

The purpose of this book is to attempt to fill the gap between studies of defence and studies of the city. It has, therefore, two complementary themes. First, that cities have fulfilled a large number of varied defence functions for which their distinctively urban characteristics have particularly suited them, and that these functions have in turn played an important role in the location, morphological form and functioning of urban places. Second, that cities, as distinctive environments, have played an important role within defence both strategically and tactically. Although both these themes are expressed above in the past tense, the phrase 'and continue to do so now and in the foreseeable future' could be added in each case.

Once this interface between defence and the city is examined, the difficulty is not so much to find instances of such interaction but to structure the many, varied and important ways in which the two have become intertwined. The absence of an existing structure for approaching either 'the role of defence in urban geography' or 'the role of cities in defence' made it necessary to construct both. This book was structured quite simply by considering the various ways in which defence and cities had interacted and by grouping each important set of interactions into chapters. The result is that the book begins with an account of the most visually obvious and well-researched defence function, urban fortification (Chapter 2), then broadens the discussion to include a range of other types of 'defence town' (Chapter 3). The focus of attention then shifts from external to internal defence in the 'insurgent city' (Chapter 4), and then to the urban environment as battlefield (Chapter 5). The battlefield theme is continued in another guise in a consideration of the city as hostage to aerial bombardment (Chapter 6). Finally, the wide variety of ways in which redundant defence works and facilities are re-used in the service of quite different urban functions is the subject of Chapter 7, culminating in their transformation into 'heritage', the raw material for a new activity (Chapter 8).

Such an approach is open to the objection that other topics could be included instead of, or as well as, those handled. However, each of the topics covered demonstrate different aspects of the three main types of interaction between defence and cities. These can be disentangled as follows.

Defence and urban location

At the most basic level, defence considerations on either the strategic or local tactical scales are a factor in the choice of urban location, either because the defendability of sites is itself an important consideration for local defence or because the settlement is intended to serve some broader strategic defensive purpose. Both of these aspects have been widely researched by urban historians and historical geographers, and numerous examples of the first are to be found in the uses of the defence potentials of local sites in Chapter 2 on the fortified city, and of the second in Chapter 3 on the strategically planned defensive urban system.

Two, general, regional-scale examples illustrate something of the extent of the pervasiveness of this factor in quite different environments at quite different historical periods. On the regional scale, spatial variations in security may be a determining factor in the intensity and pattern of settlement. The difficulty is that traders and invaders are likely to be attracted by the same set of conditions. Thus zones of insecurity are also likely to be zones of commercial prosperity.

Figure 1.1, which shows the example of the Pamphylian coast, is typical of much of the long history of urban development in many such areas on the Mediterranean coast. Good harbours on long-standing busy trade routes, along a fertile coastal plain, provide an environment both attractive to settlement and equally vulnerable to seaborne attack. Consequently, there has been a physical oscillation of settlement over some 2,000 years between

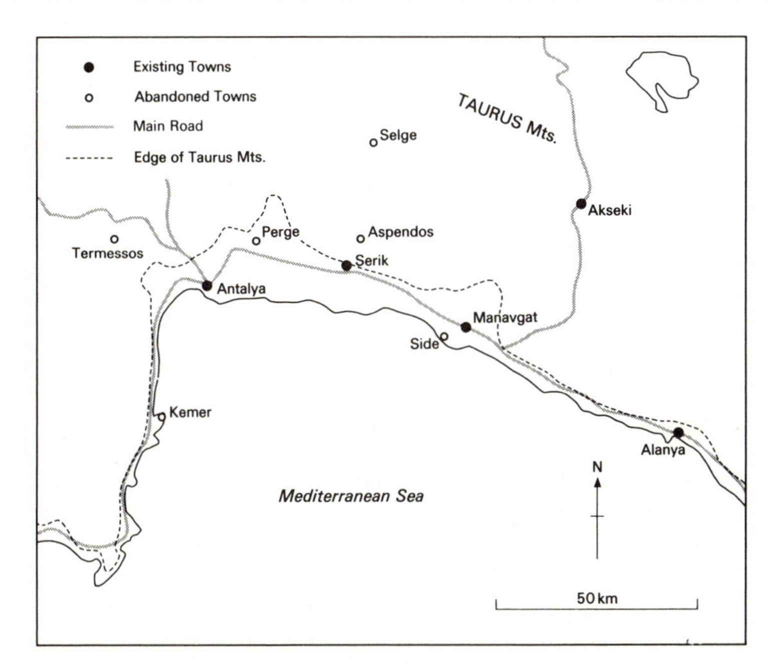

Figure 1.1 Settlement on the Pamphylian coast

the commercially profitable, but vulnerable coastal settlements (such as Antalya or Side) and the more defendable, but less accessible, inland foothill sites on the slopes of the Taurus Mountains (such as Perge, Temessos and Aspendos), as a balance is maintained between economic opportunity and security in response to the wider political and military situation.

A more recent, and North American example, is the search for a suitable location for an administrative capital for Canada, in the context of a perceived military threat from the United States (see Table 1.1). The sites on Lake Ontario and (especially) the Niagara Peninsula had proved vulnerable to attack from both land and water, while the St Lawrence river was capable of artillery interdiction, through much of its length, from the United States bank. Thus, ultimately, the choice fell upon inland sites, where distance from the US border provided protection but at the cost of a sacrifice of accessibility. The eventual selection of the lumber settlement of Bytown on the Ottawa river was aided by the completion in 1832 of a fortified canal (the Rideau), which linked Upper and Lower Canada, bypassing the vulnerable St Lawrence. Principally, therefore, defence considerations account for both the selection of the federal capital and also for much of the pattern of settlement in the defensive zone of eastern Ontario.

Table 1.1 The selection of a Canadian capital

Towns considered	*Location*	*Functioned as capital*
Niagara on the Lake, Ontario	Lake	—
Burlington, Ontario	Lake	—
Kingston, Ontario	Lake	1841–4
Toronto (York), Ontario	Lake	1849–51, 1855–9
Montreal, Quebec	St Lawrence	1844–9
Quebec City, Quebec	St Lawrence	1851–5, 1859–65
London, Ontario	Inland	1791–3
Simcoe, Ontario	Inland	1816
Peterborough, Ontario	Inland	1819
Bytown (Ottawa), Ontario	Inland	1865–present

Defence and urban form

A second broad area of interaction is that between defence and urban form. Some aspects of this have been so visually obvious as to have attracted considerable academic attention (summarized in Chapter 2), in particular the various effects of urban fortifications upon both the general patterns and particular morphological details of fortified cities, as well as the considerable impacts of the dismantling and decommissioning of defence works (see Chapter 7).

Much less lavish attention, however, has been paid to other aspects of the same relationship, especially the destructive effects of military operations on cities (see Chapters 5 and 6), and, more constructively, the re-use of defence works for other purposes, not least the creation and exploitation of 'defence heritage' (see Chapter 8). The reverse relationship (i.e. the effects of urban form upon the conduct of military or internal security operations) has curiously received very little attention. The defence–urban form relationship is the subject of Chapter 5, and the urban form–defence relationship, in various guises, is considered in Chapter 4.

Defence and urban function

The difficulty here is not in describing occasions where defence requirements operate as urban functions, so much as distinguishing between the many different types of defence needs, and placing defence within the context of the multifunctional city. The needs of defence tend to be concentrated in space and episodic in time. The account of the fortification of cities (Chapter 2), for example, shows that although defence was almost always a consideration, only occasionally at particular times and places was it of critical importance. Similarly, various defence functions can become a dominant feature of a city's economy, society or politics and create a recognizable 'defence town' (Chapter 3), or even 'defence heritage town' (Chapter 8), but these are special cases rather than more general archetypes. Two further dimensions need a brief justification; those of time and space.

This is not intended to be a historical geography, but an account of how defence and the city are related. The past, the present or the future may provide cases or scenarios illustrating aspects of this relationship, but inevitably the long history of conflict is the richest vein of such examples. History is also used in another way, that of chronology as an explanation of contemporary patterns. The evolution of fortification systems, or of attitudes towards urban insurgency, for example, require an account of change over time as explanation of the present.

There is a distinct spatial bias in the selection of examples and case studies to illustrate the topics discussed. Most are drawn from Europe, with a distinct bias towards north-western Europe, and with a sporadic reference to the spheres of European influence beyond, as the rest of the world became drawn into European rivalries and disputes. This Eurocentricity can be justified quite simply. Europe has, since the end of the Pax Romana, been the world's most belligerent continent. It has been rent by an almost continuous series of dynastic, religious, and national rivalries which for the past 500 years it has exported to every other continent of the globe.

The brief, precarious peace that has ensued since Europeans emerged from the rubble of two world wars – that were, in practice, largely European wars – is notable for its uniqueness in European history. Yet, even during this peace, preparations have been actively undertaken for a third world war, which once again would be fought out across European battlefields. It is the European cities and their citizens that stand nuclear hostage for the continuance of this peace, as it is European cities that have borne the brunt of the urban terrorism offensive since 1970, with 33 per cent of total terrorist incidents, compared with Africa's 24 per cent and Latin America's 22 per cent (Livingston 1978). It is European armies that since 1945 have fought most of the 'savage wars of peace' across the world. Such a world league table of military intervention is headed by Britain and France which have fought twenty-six and thirteen external wars respectively since 1945. Finally, it is therefore not surprising that the transformation of this long experience into defence heritage and its use, whether in military tourism or in national or regional images, should be most fully developed in this most warlike and dangerous of continents. Thus the study of war is at home in Europe, and the extent and long history of the continent's urbanization makes a Eurocentric approach to urban warfare inevitable.

Chapter 2

The fortified city

ORIGINS AND NEEDS

Fortifications can be defined as the deliberate erection of physical structures intended to provide a military advantage to a defender and impede, or otherwise disadvantage, an attacker. Natural features such as relatively high ground, or favourably located water, vegetation obstacles and the like, may in themselves endow a site with the characteristics of a 'natural fortress', but the idea of fortifications implies a series of deliberate decisions. These relate the amount of resources needed to permit the defence of a particular geographical location to the strategic and tactical benefits of doing so. The result is a technical solution which matches the resources available for defence to those of a potential enemy. Fortifications are therefore 'weapon systems' whose principal purpose is to compel an attacker to expend more time or resources on their capture than is expended on their defence. In that sense, both the rifleman's individual 'foxhole' and the rock plateau of Carcassonne – surrounded by 3-metre-thick and 1,100-metre-long walls, with twenty-six circuit towers (Salch 1978) – are fortifications, designed to allow ground to be held by a weaker defence against a stronger attack.

Given the various important defence roles that cities have always possessed, as outlined earlier, it is not surprising that the idea of fortifying cities is as old as the idea of the city itself. This is not the place to repeat the various theories of urban genesis but it does seem reasonable that the basic human daily need for sleep and shelter, and the longer-term rhythms of child rearing and seasonal food storage, made necessary the territorial defence of sedentary settlements; the coming together of people in large groups gave not only a security in numbers but also allowed the division of labour so that the strong could defend the weak. Improving the natural defences of such a settlement – with a ditch or a rampart of accessible materials such as earth or wood – seems such a short logical step that it must have been taken almost instantaneously with the decision to settle rather than seek security in flight or concealment. Certainly, archeological evidence of the earliest cities, more often than not, shows a fortified settlement. In Palestine, Jericho was walled from around 6500 BC, but despite an 8-metre-wide ditch and a stone wall with circular towers, it miraculously fell to Joshua around 1400 BC. The Sumerian cities were walled, the most impressive being Uruk with a 9.5 kilometre wall between 4 and 5 metres thick (Whitehouse 1977). The Assyrians had a particular reputation between 1300 and 700 BC as masters of siege warfare (Humble 1980), while Nebuchadnezzar's Babylon had the sobriquet 'strongest city in the world'. For the earliest Hellenic cities, we have Homer's evidence of the walls and 'topless towers of Illium' and Schliemann's (1880) description of the excavation of the 4–10-metre-high and 5-metre-thick fortifications of Mycenae. By 200 BC a school of military architecture had been established on Rhodes (Anderson 1984).

Although the focus of attention of this chapter is largely on Europe, fortified cities have not been confined to that continent or the neighbouring Mediterranean world. In terms of world-wide distribution, the city wall is a dominant feature in the indigenous urban traditions of China (Loewe 1968), the Indus Valley (Whitehouse 1977) and sub-Saharan Africa, as well as being exported to the European colonies in the Americas and Asia. The ubiquity is such that the wall becomes in many cultures essential to the definition of a place as city and the very symbol of urbanism itself.

Three of the most influential contributors to the debate on the historical origins and purpose of the western city – Henri Pirenne, Max Weber and Lewis Mumford – have supported these empirical observations with generalizations about the role of the urban defence function and its manifestation in fortifications. Weber produced the 'garrison theory' of urban origin (1958) which gave a historical primacy to defence needs and reduced many urban functions to servicing those needs. Mumford's more multifunctional view (1961) saw the three most visible elements in the urban scene, at least as far as the western city is concerned – namely, the market place, the town hall and the castle – as physical symbols of three interdependent functions. The exchange function of the market trader depended upon the regulation of order provided by law and administration, which in turn could only operate within a secure environment: the wealth generated must be defended from external attack. Similarly, the fortifications required to provide such a defence needed an ordered administration for their construction, maintenance and manning, as well as a regular supply of wealth to pay for these. The three functions of trade, government and defence are seen as mutually necessary and supporting, and the fortified city becomes an intrinsic part of the merchants' city or the officials' city rather than a separate special type of urban settlement.

Not all the different defence roles of cities (as outlined in Chapter 1) necessitated the city being fortified to the same degree, or at all. Defence against internal insurrection, the protection of a garrison against surprise attack, the defence of a town for its own sake (or as one part of a wider regional or national defensive system) all have different requirements. The fortifications of most Roman towns in the heyday of its empire were intended to do little more than legally demarcate territory, help the exercise of some police functions and provide time for a garrison to deploy in the face of surprise attack. The simple ditch, low spoil earthwork and wooden palisade of Caerwent, or the strikingly similar construction – although separated in time by 1,500 years and in space by the Atlantic Ocean – of the first British defences of Halifax, Nova Scotia (Figure 2.1), were both cheap but adequate responses to a technologically inferior enemy.

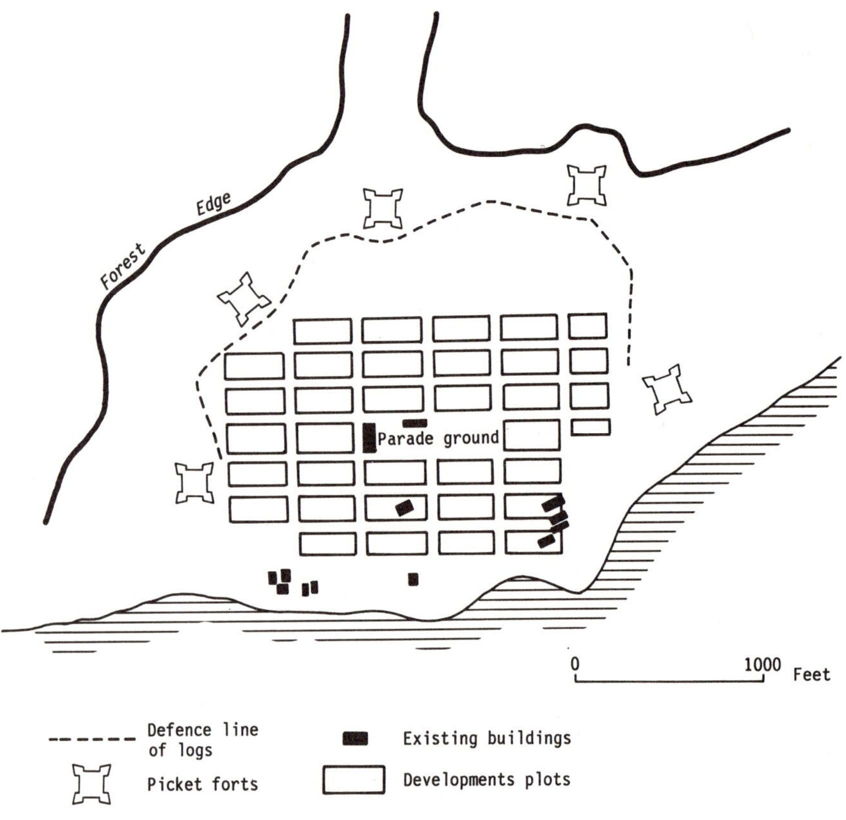

Figure 2.1 The fortifications of Halifax (Nova Scotia) 1750

Even where such discrepancies in the level of military technology or organization did not exist, the need for fortifications could be viewed quite differently in the light of different military strategies. One pair of contrasting extremes in this respect is illustrated by what can be termed the Spartan (as opposed to the Athenian) reactions deriving from the fifth-century BC Greek city-states. The Spartan reliance on the invincibility of a field army rendered fortification of the city both unnecessary and also a sign of a despised military inadequacy, while Athenian reliance on naval strength and control of colonial sea routes resulted in the building of the long walls

(476–472 BC) behind which the naval base, and the resources needed to man and supply it, could be protected.

The advantages of fortifications are not difficult to appreciate. In military terms they simply raise the ratio of attack-to-defence effort needed to capture ground, rendering it more difficult, costly or not worth the effort of conquest. A superiority in numbers of between four and seven to one was generally required to attack successfully an even moderately well-defended, seventeenth-century fortress (Creveld 1977), and such success might still take months to realize and to require specialized equipment.

However, the disadvantages to the defenders must also be appreciated. Fortifications imposed direct costs on the city for their construction and continuous maintenance. The strain on municipal finances often amounted to more than half of all city revenues in the average walled city of the fifteenth century (Morris 1972). Local manpower as well as money was required, even when the latter was supplied by the central government. The repair of the defences of s'Hertogenbosch in The Netherlands in 1629, for example, was undertaken by 17,000 men, raised by levying the labour of every sixth male adult, not only in the city but in the surrounding region. Such an imposition was not particularly exceptional. There were also the indirect costs to the urban economy through the restrictions that the physical presence of fortifications imposed on movement, land use and urban expansion. In addition, the consequences to the fortified city of its defence – whether successful or not – may have been more damaging than the surrender of an 'open' city. All these effects will be considered at length later. To the military, fortifications also impose costs by inflexibly and publically committing resources, which can not easily be redeployed, to a particular method of waging battle. Fortifications are not only intrinsically defensive but impose a particular defensive strategy which carries with it the danger to morale and self-confidence of admitting the superiority of the enemy in the open field. There is what might be termed the 'Torres Vedras' strategy – after Wellington's successful neutralization of French field superiority by using a tactical withdrawal into fortified positions in front of Lisbon between Autumn 1810 and Spring 1811 while retaining a strategic initiative. This can, however, easily become a 'Maginot' mentality – where fixed defences merely endow an attacker with unimpeded strategic mobility and a defender with an admission of inferiority.

The incidence of urban fortification has varied over space as well as through time in response to conditions both internal and external to the city. The former depended principally on the town's economic and organizational capacity to defend itself, and the latter on the extent of the external threat to it. Obviously towns located in regions of high risk were more likely to be fortified. This situation may be the result of a location in disputed border regions whose allegiance was regularly contested by neighbouring powers. One of many such examples in Europe is the middle-Rhine rift valley of Alsace – whose favourable position on a set of trade routes encouraged the growth of trading towns such as Mulhouse, Colmar, Selestat and Munster but whose similarly central position between expansionist military powers in Hapsburg Germany and Bourbon France led to regular invasion. Consequently, not only were the towns themselves walled, they also proved capable of using their wealth and position to maintain a precarious independence as 'imperial free cities' in the thirteenth and fourteenth centuries. They even concluded the defensive alliance of the Decapolis in 1379 in an effort at collective security. This proved effective against the Hapsburgs in the sixteenth century but failed to prevent conquest by France in 1673 (L'Hullier 1955).

A different sort of threat was posed by an external environment of widespread small scale lawlessness rather than organized military aggression. In thirteenth-century England (by no means the most lawless example) the situation outside the towns was such that: 'even outside periods of civil war the misdeeds of certain barons and their retainers ... were at times so frequent and serious that parts of the country served to be in a state of war' (Jusserand 1889). The Statute of Winchester (1285) attempted to legislate for the safety of inter-urban roads by demanding the removal of cover for ambush for at least 200 feet on either side. Fortifications in such regions were necessary for simple police rather than military precautions for the protection of property and individuals. They could be of the simplest construction but, in the regions of most chronic disorder (such as the Anglo-Scottish borderlands between the thirteenth and sixteenth centuries or the East German March in the fifteenth to seventeenth centuries), the need for fortifications would be felt right through the settlement hierarchy – from towns and villages even to hamlets and individual farms.

Conversely, the removal of the need for fortification by the imposition of a rule of centralized law, and the enforcement of order on the countryside, encouraged their early abandonment. Few towns in the Egypt of the Pharaohs were ever walled. The most obvious more recent example of this in Europe is England, where the almost complete freedom from foreign invasion, together with the early imposition of central authority, prompted most cities to abandon their walls early in the medieval period – or, in the case of many towns, not even to begin their construction. This situation was most clearly demonstrated in the feeble (and, by continental standards, amateurish) attempts to fortify the towns when that central order broke down during the English Civil War of 1642–9. This absence of the constraints of encircling walls has proved in turn to be a major element in the morphological contrast between England (with the exception of its Welsh and Scottish borderlands, where towns lacked this restriction on physical expansion) and continental Europe.

If fortified cities were a response to an absence of centralized authority, their presence could be viewed as an obstacle to the establishment of such an authority or a threat to it once established. The right to fortify was often a jealously guarded monopoly of the central government, to be granted grudgingly to cities as a reward for service or in return for favours. Most town charters establish the three essential urban rights – namely, to a market, to self-government and to a wall. The wall becomes the visible symbol of municipal independence, as the castle is the symbol of its subjection to feudal power.

The attempt of the monarchy to enforce its control in France in the seventeenth century, for example, used the demand for compulsory dismantling of fortifications as an instrument of such control by removing the possibility of local urban dissent from royal authority. The resistance to the dismantlement edict of Louis XIII, and his minister Richelieu, by the city of La Rochelle (one of around 100 fortified Huguenot centres) was a recognition that the preservation of a separate religious identity was dependent upon the preservation of a degree of independence conferred by a fortified city, a '*place de sureté*' as had been guaranteed by the 1598 Edict of Nantes. The enforcement of central government control led in 1628 to one of the bitterest fought sieges of the period (Salch 1978), if only because it was powered on both sides by absolute values, between which compromise (the more usual outcome of siege operations) was impossible.

Cities have been fortified for various defensive purposes as part of a wide variety of military strategies. In addition, defence systems cannot be divorced from the political and economic systems that produced them. In one society they 'represent the ethnical-political idea of the security and strength which ensured the freedom of the citizens' (Argan 1969), a visible guarantee that 'stadtluft macht frei', while in another the same fortifications may represent economic oppression and political subordination. This chapter will first outline the chronology of the development of urban fortifications in response to the evolution of weapons technology, military organization and defence policy. It will then be possible to consider some of the relationships between fortifications, on the one side and both the development of urban form and the functioning of cities on the other.

THE DEVELOPMENT OF URBAN FORTIFICATIONS

The difficulties of chronology

The development of the applied science of fortification is intimately related to the history of weapons technology. This 'continuous dialogue' (Rolf and Saai 1986) is in a constant state of action (as a new weapon is invented or an old one applied in a new way) and reaction (as a method of defence against it is sought). However, what could be portrayed as a simple narrative of discovery, the use of an improved method of assault, and the successful improvement of defence to counter it, is complicated by a number of factors.

As with all innovations, the adoption of new techniques in military science is not instantaneous but occurs slowly over time and unevenly through space. There are aspects of the practice of war and its practitioners that render it likely to be particularly conservative in its abandonment of the old and adoption of the new. This is a result, in part, of the intrinsic difficulty of the empirical testing of military ideas. Every military action involves so many variables that are to such a large extent unique to that situation, and chance is so important an element in determining the outcome, that a series of experimental results can rarely be obtained as proof of the success of new methods and the obsolescence of the old. A strong reliance on the tradition of the 'tried and tested' is almost inevitable.

In addition, the cost and physical durability of fortifications discourages the rapid writing off of the investment and its modernization. Suleiman's sixteenth-century walls of Jerusalem were still in use in the 1850s (Cohen 1977). Adaptation and re-use of the existing structures is more likely than new building, and fortifications developed as a response to a particular attack technology may be pressed back into service many years later to meet a quite different assault. One of the most dramatic illustrations of this was the siege in 1453 of medieval Europe's largest and most renowned fortified city, Constantinople. A series of fifth-century walls, hurriedly repaired and inadequately manned, resisted for 14 months the most successful (and advanced in terms of its leadership and artillery equipment) fifteenth-century army. More recently, and with an even larger time discrepancy, machine-gun bunkers to repel the expected invasion of Britain in 1940 were built into the Roman walls of Richborough, which had been originally constructed to repel attacks in the third century (Kightly 1980). The costs and difficulties of changing fortification systems made cities reluctant to do so, unless the threat was seen as real and immediate enough to warrant such effort. The spatial and temporal unevenness in the adoption of military innovation meant that 'old-fashioned' fortifications might be perfectly adequate in meeting many threats in many parts of the world, long after they were obsolescent in others. Wooden stockades around seventeenth-century Siberian (Sumner 1944) or North American (Kerr 1960) settlements co-existed in the world of Vauban's complicated architectural masterpieces.

As towns have generally survived through many centuries of change in military technology and have responded with fortifications according to their varying needs and capacities, the result is that most are palimpsets of 1,000 years of defensive architecture. Conversely, very few are clear textbook examples of any one period, and these cases – almost by definition – can be regarded as atypical in their experiences rather than representative.

Thus any attempt to place particular dates against the chronology of weapons development, in search of key moments in history, is particularly suspect and likely to be relevant only for specific regions. The following grouping into four categories (i.e. pre-gunpowder fortifications; early gunpowder-artillery fortifications; mature gunpowder-artillery fortifications; and the last fortified cities) is therefore based on the similarities in the urban reactions to particular military situations rather than on sharply distinguished divisions of time and space.

Pre-gunpowder fortifications

This somewhat negative designation is being used to cover the long period from the first urban fortifications to the effective use of gunpowder-propelled artillery in siege warfare in the fifteenth century.

The principles and most of the practice, however, varied little throughout this long period. The purpose of fortification was simply the prevention of escallade by providing a barrier to the enemy charge. A ditch, earth-bank or wooden, brick or masonry obstruction (or a combination of all three) formed the unchanged basis of the design. Protection from the enemies' short-range, hand-propelled weapons and an elevated firing platform for discharging one's own weapons was an additional advantage. The distinction between a fortified iron-age, hill-top settlement – with its *muros gallicus*, i.e. wooden beams bound together and packed with earth (Johnson 1978) – and the most complex piece of fourteenth-century military architecture, is only one of elaboration on the basic design.

An important dilemma lay at the heart of ancient and medieval warfare. On the one hand, cities were difficult to avoid, as they commanded communications and were important reserves of material resources as well as the symbolic prizes of victory. 'The urban centres, not the castles, were the true masters of medieval space' (Contamine 1980). On the other hand, a seriously defended fortified city was extremely difficult to take by assault during most of this period. Much of medieval war therefore revolved around a series of urban sieges, which were frequently long drawn out, often ineffective, but formed the most impressive military episodes of the ancient and medieval worlds. Two examples of the many possible demonstrate this point. The Athenian siege of the double walls of Syracuse (415–412 BC) became a 'siege within a siege' as attackers and defenders changed status, and led directly to Athenian defeat in the Peloponnesian war. The siege of Acre lasted from June 1189 to July 1191 and involved an enormous concentration of international effort before falling to Crusader assault – although its subsequent reconquest in 1290 took a siege of just 6 weeks.

The methods of assault open to the attacker are fairly obvious, and, unfortunately for him, were equally obvious to the defender – which is a further reason why very few medieval towns actually fell to assault (de Vlerk 1983), and the few that did have a rarity causing them to be disproportionately remembered in history. Non-assault techniques of capturing the city, which would usually be preferred even though they could rebound on the besieger as much as the besieged, included starvation, infectious disease and psychological pressure. These would normally be used if time was no object, leaving treachery and threats of reprisal if urgency was required.

> To capture a town before the resources of the surrounding country gave out was a cardinal problem of warfare, and to solve it the curious (if logical) arrangement under which the terms obtained by the garrison stood in inverse proportion to the length of its resistence, was instituted.
> (Creveld 1977: 28)

If assault was necessary, however, then a number of methods of coming to hand-to-hand combat with the defenders were available. Ditches and moats could be bridged or filled, and the defenders' advantages of cover and elevation neutralized by approach trenches, portable screens and movable towers. The fortifications themselves could be breached by sapping, if the sub-soil was suitable, or by artillery.

This last possibility needs some elaboration, as the discharge of missiles, either muscle or mechanically powered, long pre-dates the use of gunpowder. The *onager* of the ancient world or the *trebuchet* (a product of the medieval crusades) which were both balance- or tension-powered launchers of shaped stone missiles, were used to break down gates and batter down walls by repeated hits on their points of weakness. The siege of Jerusalem by Titus in AD 70 was conducted with the help of onagers reportedly capable of firing a 20-kilogram stone some 400 metres (Grant 1974), although its impact velocity at that range cannot have been high. Technical improvements in the period 1180–1220 (Contamine 1980), perhaps as a result of crusading experience, enabled a 100-kilogram stone to be propelled on a high trajectory some 150 metres with reasonable accuracy, as was demonstrated at the siege of Tournai (1340). Such equipment was still playing a major role during the unsuccessful siege of Rhodes (1480).

An alternative use of powered missiles, equally as traditional, was what would today be termed 'suppression fire' (i.e. sweeping the defenders from the walls in order to cover an assault). Even the perfection of such a simple missile weapon as the sling, around 250 BC, resulted in a traceable reaction in the design of hill forts (Brice 1984). Similarly, a clear account of this use of suppression fire is given in the attack in 36 BC on a rebellious Han city in north-central China (Loewe 1968), where crossbow fire drove the defenders from the towers allowing the attackers to approach the wooden walls, set fire to them, enter and take the city. Similarly, archaeological evidence suggests that the assault of the second legion ('Augusta') on the important, iron-age, walled settlement of Maiden Castle, during the conquest of Britain, in AD 43 was achieved by a standard Roman technique of accurate *ballista* (i.e. relatively high-velocity bolt fire, operating at 200–400 metres, and thus beyond the range of hand-held weapons) driving off the defenders from the ramparts, allowing the gates to be stormed by infantry.

These methods of attack could be resisted by counter-attacks, sorties of

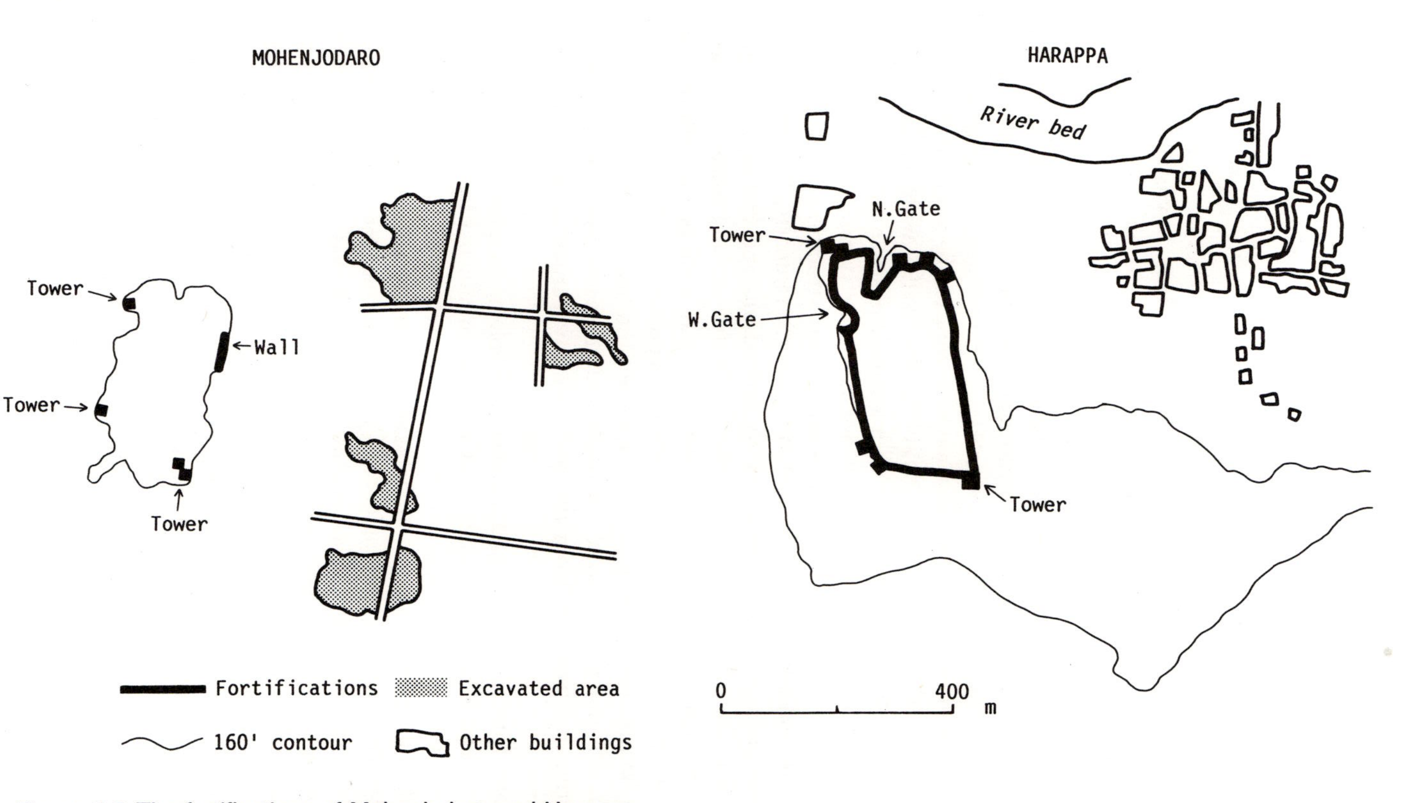

Figure 2.2 The fortifications of Mohenjodaro and Harappa

the garrison to destroy siege works or seize artillery, or by counter-mining. Throughout this long period, progress in the art of siege warfare went hand in hand with similar progress in the art of fortification. Generally, little could be done about the choice of site of towns – unlike castles, whose natural advantages for fortification or mining had to be largely accepted. It was possible to incorporate natural features to strengthen a defence, or to reduce the amount of construction required. Many towns, for example, incorporated rivers into their fortification systems as obstacles: both York and Utrecht are notable large-scale examples where at least one flank of the town defence rests upon a broad river.

Improvements could be made to the curtain wall itself. The height and strength of walls was largely dependent on the choice of building material. Brick and, especially, stone allow greater height for less width than earth or sunbaked mud, and a more vertical angle of slope can be achieved. There was also the costly alternative of doubling the walls, with an inner wall overlooking the outer, providing both a double obstacle and an intervening 'killing ground'. The two weakest points in any system were the corner angles and the gates. The corners were both structurally vulnerable to battering and difficult to cover with fire from the ramparts. The result, from earliest times (see the examples of Mohenjo-Daro and Harappa in Figure 2.2), was the strengthening of corners by buttresses, and towers for use as anti-personnel weapon-launching platforms, as well as the reduction in their number and sharpness of angle as far as possible. In this respect a circle is to be preferred to a square but although the nearer to a circular form, the fewer and more obtuse the angles, at the same time the amount of covering fire available from the walls for any particular point inevitably declines.

Gates formed a serious and unsolvable problem. As an entry point to the city, it was clearly a weakness that must be minimized by restricting the number and ease of use. Equally, however, the economic purpose of the city – and its defence – depended upon ease of communication with the outside world. There was thus an understandable and permanent source of friction between military and civilian interests. The result was to concentrate the attention of attackers and defenders alike upon this feature. The simplest solution was to overlap the walls so that two sharp changes of direction were necessary in order to enter through a narrow gap parallel to the walls and thus covered by them, and allowing no direct assault on the gate itself. Such an arrangement can be found in the traditional African village, reproduced in stone around 1000–1300 AD in Zimbabwe (Figure 2.3); and today this still forms the standard method of producing a road-block. The strengthening of gate defences with towers and outworks evolved into the barbican – a system of double, or even on occasion triple, gates connected by a defendable overlooked space. The walled city of Aigues-Mortes, for example, was constructed in 1246 as a royal military base to support crusading operations,

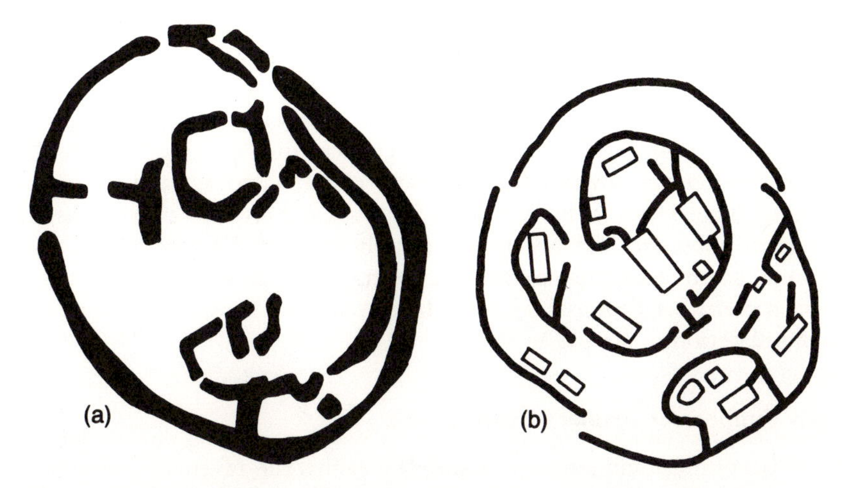

Figure 2.3 The fortifications of: (a) Zimbabwe; (b) a traditional Zulu village

and the main gate defences are so extensive as to dominate and overlook the entire structure (Salch 1978).

The evolution of these ideas over time is best examined in the case of a single, major, well-documented example. Utrecht occupied a central location in the medieval Low Countries and, although originally a Roman foundation, received the right to construct walls in the city charter of 1122. A 4-kilometre circuit – which provided ample room for future expansion and was, in fact, not outgrown for 400 years – enclosing a remarkably large area for the time, was constructed (including ditch, except along the river bend, stone wall and wooden palisade). Nine square towers were added to reinforce the wall during the thirteenth century and once brick became cheaper in the fourteenth century it was used to strengthen and raise the wall to around 5 metres, with a thickness of between 0.8 and 1.5 metres. Four land gates (all of which were barbicans) and two water gates (allowing access to the canal system) were constructed.

The effectiveness of fortifications, however, depended not only on their architectural design but equally on the efficiency of their continuous maintenance and their manning. Fortunately, the municipal records of these activities in Utrecht have survived and been published (Vlerk 1983). The fortifications were a municipal responsibility overseen by a local official, the *Schutmeester*, who was part municipal accountant, part public works inspector. The walls were divided into sections, each of which included a tower and a stretch of the ditch, and each section was the responsibility of one of the town's twenty-one guilds. The guild was responsible for routine maintenance and small repairs. In practice, one of the most onerous tasks concerned the water-filled ditch, the *gracht*, which needed constant dredging, rubbish removal, bank clearance and ice breaking in winter. Only major repairs and the care of the main gates were the responsibility of the town council as a whole. In addition, the guilds were charged with the routine manning of the walls, especially the provision of the night-watch, based on the watch houses (*wachthuisjes*) spaced at intervals along the wall. In recompense for these efforts, the guilds were permitted to exploit their 'wall' for profit wherever possible. Towers were rented for storage and crafts, the banks of the gracht were used for growing grass and for such activities as the drying of linen, and the water course itself for land drainage and sewage disposal. Despite all this continuous effort, or perhaps because of it, the town was neither captured by assault nor seriously threatened during the medieval period.

Early gunpowder-artillery fortifications

There is a popular and attractively simple idea that the advent of gunpowder caused a sudden, fundamental reassessment in fortification design by rendering the existing technology obsolete and tilting the balance of forces decisively in favour of the attacker. Although this sort of revolution, as the result of sudden advances in weapon technology, is not unusual in military science, there is considerable doubt that this is what occurred in this case for three reasons. The effective use of gunpowder-propelled weapons in attacks on cities was perfected over a long experimental period of almost two centuries; the new artillery did not at first convey more advantages to the attacker than the defender of fortifications; and fortification design was able to adapt over the long transition period to meet, at least partially, the new threats to it. For many centuries after its introduction, 'the invention of gunpowder in no way diminished the role of the fortified city as a stronghold' (Brice 1984: 104). It did, however, alter the nature of those fortifications and their influence upon war as a whole.

Neither gunpowder nor artillery were inventions of the fourteenth century. As has been mentioned earlier, stone-firing artillery machines, designed to batter down defences, had a thousand-year history of use and had been perfected by the thirteenth century. Experiments with gunpowder were designed to increase the launch velocity of existing pre-gunpowder artillery to achieve two improvements. First, to increase the range over which it could be used beyond the existing 100–150 metres, so as to allow artillery to be sited outside the range of defending missiles, and, second, to increase both its hitting power and accuracy. These last two were essential if artillery was to be effective against town defences – which could only be damaged by repeated hits from stone projectiles on the same place. In the siege of Rhodes (1480) some 3,500 balls were fired before an exploitable breach in the fortifications was made (Banks 1973).

It is not clear when gunpowder was first used to expel a projectile from a re-usable weapon for military purposes. There is a record of such an event occurring in Ghent in 1313 (Allen 1976) but even more notable is that such a novelty, whether completely original or not, was within a generation being applied to attacking urban fortifications. However, even though gunpowder was being used in urban sieges quite early in the fourteenth century, such as the siege of Friuli (1341) or Calais (1346–7), it had little more than the psychological effect of novelty until the 1428/9 siege of Orleans by the English.

First, the technical problems of increasing the muzzle velocity concerned the nature of the propellant, as the chemical formulae for gunpowder were uncertain until the middle of the fifteenth century; in addition, the ingredients were expensive and generally impure. Second, the shaped stone projectile was unpredictable in weight and shape, which made precise calculations of propellant as well as gun elevation impossible, until the perfection of cast-iron shot around 1480. Third, increasing the muzzle velocity posed new demands on the launcher and thus new advances in metallurgy and gun-casting. Barrels were not cast in one piece until early in the sixteenth century. It is not so surprising, therefore, that the adoption of gunpowder artillery – dependent as it was on parallel developments in many fields of scientific knowledge – took so long.

Throughout most of the fourteenth and fifteenth centuries in Europe, therefore, gunpowder-propelled artillery was extremely short-range, and was both large and heavy in relation to its hitting power. The consequences of these two conditions were that the new artillery was largely immobile. The development of the wheeled gun carriage – first recorded in 1494 (Brice 1984) – and *trunnion* (i.e. the protusions on the barrel which are used both to support it on the carriage and later allowed it to be easily elevated) were thus extremely important in determining how artillery could be used. Until well into the sixteenth century such field pieces as could be manouvred with difficulty had to be positioned very close to their targets. This in turn gave the defenders two advantages: (1) the besieging artillery was very vulnerable to counter-sortie and destruction *in situ*; (2) it was easier and safer to use static artillery in the defence of the city than field artillery in attacks on it (Collier 1980).

Provision for defensive gunpowder-propelled artillery was quickly incorporated into fortification systems. Such artillery served two main purposes. One was anti-personnel fire for the protection of points of likely assault, such as gates and walls, using either small-bore weapons, or large bore weapons firing short-range, but widely spread, multiple shot of one sort or another. The loopholing of towers, especially gatehouses, was an early reaction. The other use was counter-battery fire designed to prevent attacking artillery from approaching within effective range of the walls. For this purpose raised gun platforms with wide arcs of fire were required. These *bastions* were generally strengthened, and often truncated, existing towers but it became obvious that purpose-built bastions placed in front of the wall

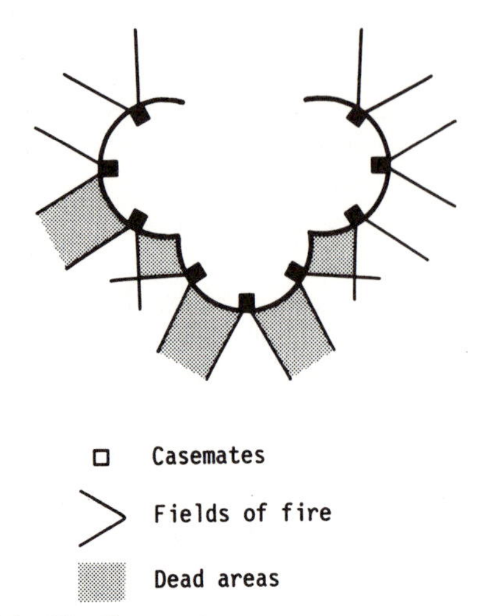

Figure 2.4 The trefoil fortification system

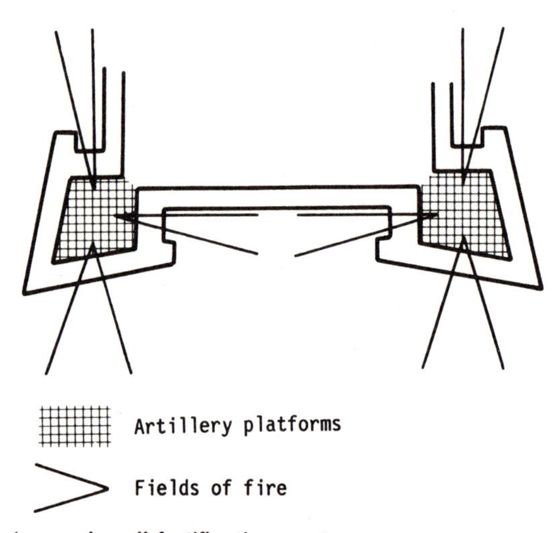

Figure 2.5 The 'arrow-head' fortification system

had two advantages: (1) by being built out from the wall they increased its distance from enemy artillery; (2) they could provide anti-personnel covering fire along the wall itself. Various bastion designs were tried. A pattern of semi-circular platforms joined in a clover leaf was the design for the series of purpose-built artillery battery emplacements erected between 1546 and 1558 to defend the coastal towns such as Deal, Tilbury, Portsmouth and Yarmouth from the threatened invasion of the Spanish Armada. This *trefoil* design (Figure 2.4) was rapidly superseded by the *arrow-head* design (see Figure 2.5), which had the advantages of eliminating blind spots and providing interlocking, 270-degree firing arcs so that a small number of guns could defend a large length of wall.

The increased accuracy and hitting power of artillery in the course of the sixteenth century was difficult to counter, and the most reliable technique was to keep the enemy guns out of range rather than find ways to withstand their fire. Nevertheless, it became clear that the resilient strength of fortifications in absorbing fire was more important than height, which merely presented an easier target. Towers and even walls tended to be reduced in size and strengthened. Towers, in particular, needed strengthening in order to bear the weight of new artillery, which could, moreover, be more easily moved if sited on the same level as the walls. The walls themselves were thickened, often being backed by shock-absorbing, packed-earth reinforcement (*retirata*). In Utrecht, for example, a gently sloping earth bank was constructed on the inside against the stone walls in 1512. In 1528 and 1544 artillery bastions were added to the walls and in 1577 five, new, earth gun emplacements (*bolwerken*) were added in front of the most vulnerable sections of wall.

Although most towns adapted their existing medieval fortification systems in this incremental way, there were some examples of entirely new fortifications being constructed, for the defence of specific important key positions, according to the theoretical principles of the newly articulated science of military architecture. Although these provide clear-cut, dramatic and (quite literally) textbook examples, as they were copied in detail from the published blueprints of the theorists, it must be remembered that most towns did not receive this attention but engaged instead in the sort of piecemeal adaption of their existing defences, as municipal finances allowed or external threats compelled.

These fortification systems of the sixteenth century, whose evolution was prompted by the increasing efficiency of cannon, were only possible as a result of two more general trends in European society. First, the development of absolutist, centralized organizations was necessary for financing and executing such works, which represented a scale of public investment beyond the funds of all but the richest individual city. The show-piece, walled cities of the sixteenth century were the result of national investment, whose objective was not merely the defence of the town within the walls, but the denial to the enemy of space, which was of critical importance in some wider, regional, national, or even continental, defensive context. In other words, these cities reflect enlargement in the scale of warfare on the European continent. Second, the invention and spread of printing and, even more fundamentally, the growth of an international intelligentsia as a market for printed material during this period, had an important influence upon military science. It not only transformed military engineering and architecture from a craft into a profession but also allowed the rapid diffusion of a body of accepted modern practice throughout the continent. For example,

Albrecht Dürer's *Discourse on Fortifications* was published in Nuremburg in 1527 and was used as a model for numerous German towns in the next 80 years (Argan, 1969), and Franchesco de Marchi's *Architettura Militare*, although published in Italy in 1565, could be found applied in examples in The Netherlands, Austria and even England within 10 years.

By the middle of the sixteenth century contemporaries demonstrated their awareness of both the extent and origin of the changes by referring to the 'new', or 'Italian', system. The combination of wealth and internecine warfare in northern and central Italy, and the concentration of both on the cities, explains why the scientific advances of the renaissance should have been coupled to the need for urban fortifications in this region. Many blueprints were constructed in the search for a military architect's Utopia. Francesco Martini (1451) multiplied and developed the corner bastions within a variety of geometrical figures. Filarete, in 1464, minimized the vulnerable corners as well as providing defensive artillery-covering fire for all the walls with no blind spots, and a town layout was designed for the rapid deployment of its garrison to threatened parts of the defences, thus making an early link between the external wall and the internal structure of the city. This 'sforzinda' was an idea rather than a town but a number of fortress towns defending key positions on the North Italian Plain were built very closely to such specifications – such as Novara and Palma Nova (Carter 1983).

The long struggle for the Low Countries – 'The Eighty Years War' (1572–1648) – occurred within a set of circumstances not dissimilar to northern Italy a generation earlier. The war was, in one important respect, unequal – as it pitted what was generally regarded as the best field army in Europe, led by some of the best commanders, against prosperous but essentially non-military urban populations. The possibility, and financial resources, to fortify the dense network of towns restored the balance.

> The strength of a late sixteenth century and early seventeenth power no longer consisted mainly in its field army; instead it lay in the fortified towns and a country liberally studded with these would even find it possible to wage war without any real field army at all.
>
> (Creveld 1977: 9)

It is not surprising, therefore, that the Low Countries were early adopters of the 'New Italian' system, which found such an enthusiastic application as to become known as the 'Old Dutch' system (Sneep *et al.* 1982).

If the medieval walled town was effectively defensive and tactical, the system of fortified towns developed in The Netherlands in this period was equally aggressive and strategic. The most influential theorist was Simon Stevin, of what was then the new University of Leiden, and the most renowned engineer was Andriaan Antonisz (1541–1620). It is notable that military engineering had become a university scientific discipline and that,

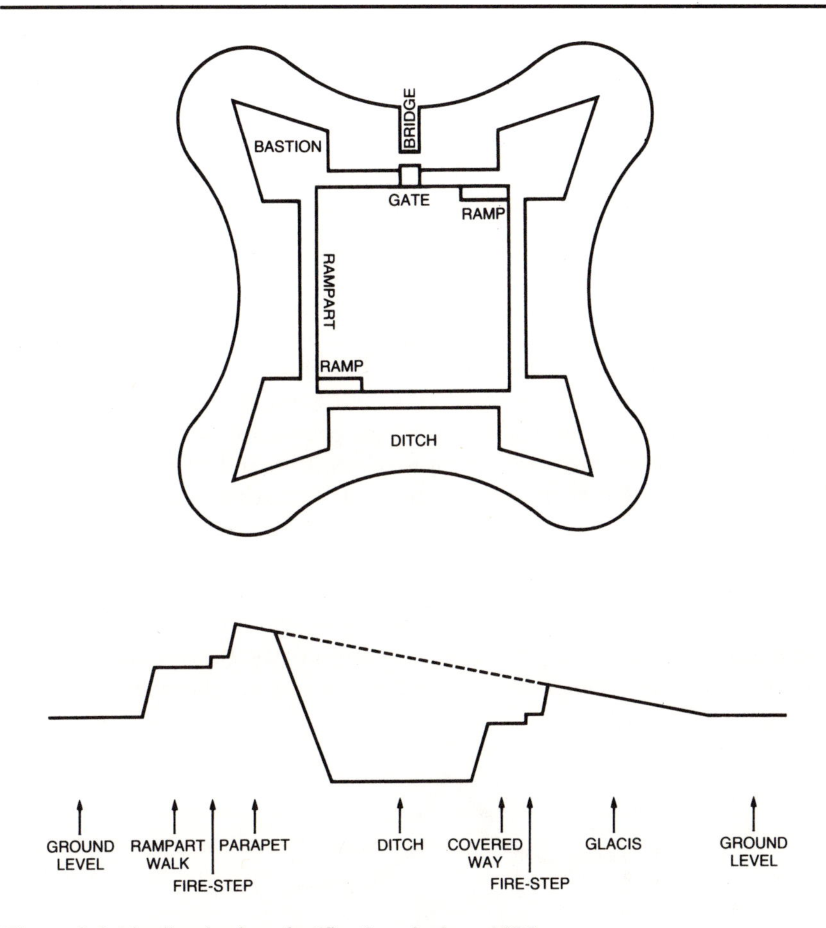

Figure 2.6 Idealized urban fortification design, 1600

for the first time in the long history of urban fortifications, the builders are named individuals.

At its simplest, the theoretically ideal plan evolved by 1600 is shown in Figure 2.6. The cross-section demonstrates the new significance of height levels designed both to minimize the wall as a target and to sweep attackers off the *glacis* with defensive fire.

Antonisz and his pupils were responsible for the refortification of many of the towns of the southern and eastern borderlands of the United Netherlands but his masterpiece is the small but geometrically perfect purpose-built fortress town of Willemstad. The importance to the United Netherlands of holding a bridgehead on the south bank of the Rhine distributaries resulted in 1585 in the design which fulfilled the requirements of Stevin, then Quarter-Master General of the states' army. It was executed almost exactly to this design, and has very largely survived unchanged to the present (see

Figure 2.7). Each of the seven walls of the polygon are covered by interlocking fire from at least two of the seven 'arrow-head' artillery bastions. The wall and surrounding water-filled gracht are pierced only twice – for a water gate and a land gate, the latter being covered by a single detached *revelin*, causing an entrant to make three abrupt changes in direction. The town was intended to serve the fortifications rather than the reverse, housing the garrison, its families, and necessary military and civilian services. The main streets are broad and straight – so as to accommodate military vehicles and, in particular, allow the rapid movement of artillery pieces between bastions.

Although the new fortification system had its origins in northern Italy and its proving ground in the Low Countries, the speed of the spread of its adoption throughout the rest of Europe and beyond was remarkable, and was aided by the internationalization of the profession of military engineer and the rise of absolutist national monarchies. In Scandinavia, for example, the emergence of Denmark as a centralized monarchy, with ambitions to dominate the Baltic, resulted in a spate of fortification construction – especially under the great builder Christian IV. He was responsible for ten new towns, and the modernization of the defences of many more between 1600 and 1650, including: Kristiania (modern Oslo); Kristiansstad; Fredericia; Gluckstad; and, in reaction to this activity, the Swedish fortification of Kalmar and Goteborg. All had the basic plan of stone-faced earth

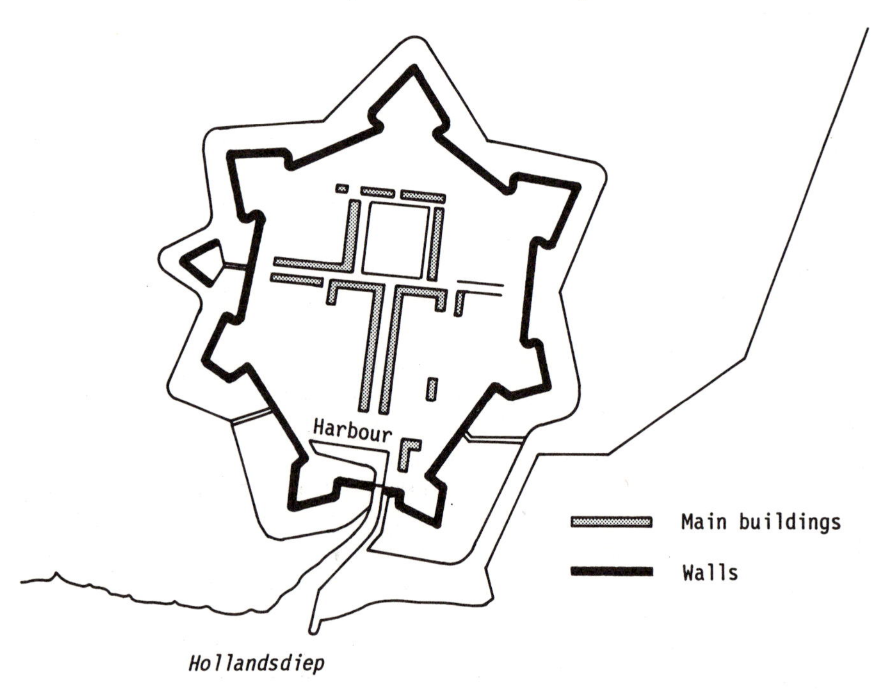

Figure 2.7 Willemstad (The Netherlands)

ramparts and arrow-head artillery bastions. However, the most notable set of fortified towns occurred along both sides of the narrow disputed waterway of the Sound, the key to domination of the Baltic. At its northern entrance stood the twin towns of Helsinger and Helsingborg, 5 kilometres apart, while at the southern end were Malmo and Copenhagen, making this the most defended stretch of water in Europe at the time. In particular, the need for a secure naval base led to the construction of what was effectively a new strongly fortified town to the south of Copenhagen, Christianshavn.

The enlargement in the spatial scale of warfare in the seventeenth century was not confined to Europe. The expansion of European trade overseas caused parts of America, Africa and Asia to be used as battlefields for European rivalries. The seventeenth century thus shows the beginnings of a world economy and its concomitant world war, and the need to fortify settlements was quickly exported. Traders needed a base for the storage and transhipment of cargoes. Such bases needed protection – not only from locals, whose level of military technological development was likely to be lower than the European, but, more important, from other European commercial rivals. The merchant was thus rapidly followed by the soldier, the administrator, and the civilian settler. In North America three-cornered Spanish, French and English rivalry led to a series of fortified settlements along the Florida and Carolina coasts. One of the earliest was St Augustine (Florida) which in 1565 was fortified with a wall (including eight artillery bastions and a star citadel) by the Spanish in 1565, the necessity for which was demonstrated by it changing allegiance four times in the following 100 years (Raisz 1964).

Further north, French and English rivalry for the fur trade of Canada led to the foundation of Quebec City in 1608. Founded for strategic reasons, it was sited on a defensible position above, and commanding, the St Lawrence routeway, and was planned as a walled city with artillery emplacements covering both the landwards approach and the river routeway. Although its description as 'North America's nearest approach to a medieval European city' (Nader 1976), is astray by a few centuries, it is true that it was walled in the seventeenth century to contemporary European standards, and became a prize of such conflicts for the next 200 years, during which the fortifications were continuously modernized.

Other Canadian cities were walled during this period (see Kerr 1960) but to a standard well below that prevailing in Europe. Montreal is an example of this 'colonial time-lag', being provided with a palisade against Iroquois Indian attack in 1685, and being walled in stone in 1717 and 1744. As can be seen in Figure 2.8, the fortifications were rudimentary, the truncated bastions suitable for short range anti-personnel weapons and no real resistance was offered against the attack of the British in 1760 or the Americans in 1775. The plan for the defence of Halifax (Nova Scotia) – drawn up soon after its foundation, as late as 1750 – shows a colonial

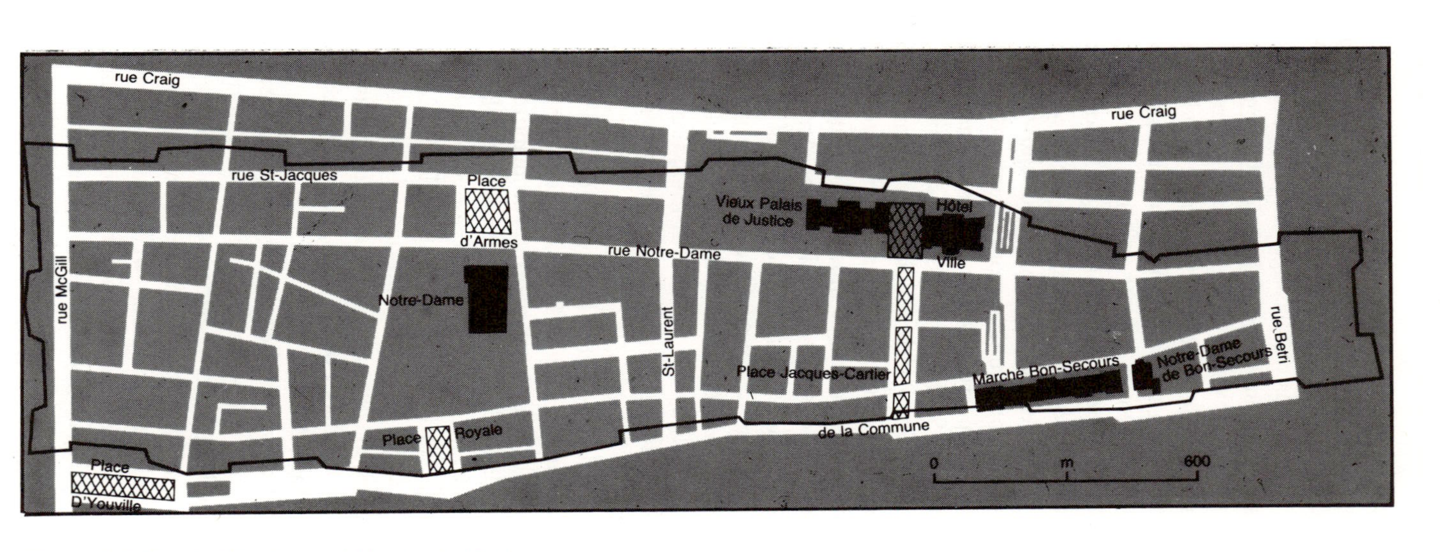

Figure 2.8 The fortifications of Montreal, 1744

simplicity (see Figure 2.1). This may have been adequate defence against Indians or dissaffected settlers but would not have been a serious obstacle to an artillery-equipped European force. A 200-metre field of fire has been cleared and the felled trees used in the main defence line. Five small 'picquet' forts, each a battery behind earth and wood bulwarks, protect the landward approaches but the assumption that the Royal Navy covers the seaward side is clear. Fortunately, this early plan was soon superseded by more substantial fortifications once war with France was imminent. More effective fortification building in North America was, in general, an eighteenth- and nineteenth-century phenomenon.

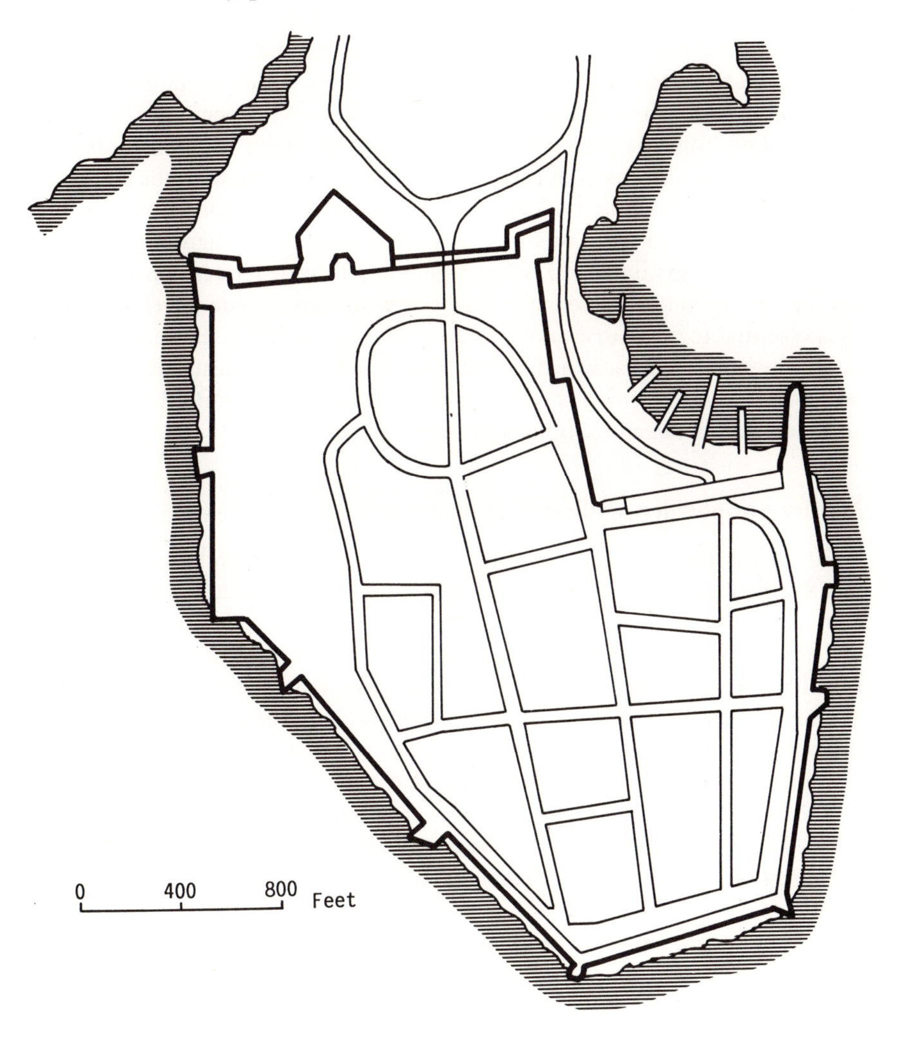

Figure 2.9 The fortifications of Galle (Sri Lanka)

The export of fortification technology outside Europe was not confined to North America. The West African coast from Senegal to Nigeria is dotted with the remains of the fortified factories of English, French, Portuguese, Danish, Dutch and German traders, some of which developed into modern cities, as also are the West Indian islands at the other end of the trade route. The Dutch towns of Sri Lanka are among the best documented examples of the European fortified town of the seventeenth century transplanted into a colonial setting, where the most recent technology of the Eighty Years War was applied to urban defence against both the native Singhalese and European rivals, especially the English (Nelson 1984). A number of such towns were constructed after 1658, or often reconstructed on the pre-existing Portuguese foundations, including the important modern cities of Colombo and Jaffna, but one of the most complete surviving examples is Galle on the south-west coast (see Figure 2.9). The peninsula site was protected on both the seaward and, even more strongly, landward sides, by low artillery bastions built of earth faced by laterite, fronted on the landward side by a 15-metre-wide ditch and a clear field of fire. A special problem of the colonial fortified town was that they were generally foreign implantations in a potentially hostile inland environment and thus dependent on supply by sea. The provision of a defendable harbour with its quayside warehouses was thus essential for the survival of the town as well as being the reason for its establishment. Such a seaward entrance breaches the fortifications, creating a point of weakness, which is compensated for by two covering bastions and an inner wall between harbour and town.

Mature gunpowder artillery fortifications

Many of the trends mentioned above, especially in the technical improvement of artillery, continued at an accelerating pace throughout the second half of the seventeenth century, which in turn led to improvements in the science of fortification that were radical enough to be known, in Northern Europe at least, as the 'New Dutch System'. The result was the building of urban fortifications that are arguably the most architecturally dramatic, technically complex and aesthetically satisfying in the history of military engineering – to the extent that the two most prolific designers of the period, who were effectively contemporaries, Sebastien de Vauban (1633–1707) and Menno de Coehoorn (1641–1704), are the two military engineers most popularly remembered in the whole history of the science.

It is not surprising that France and the Dutch Republic should have taken such an active interest in these affairs. The second half of the seventeenth century was a period of French military assertiveness and expansion under a centralized monarchy, not least in the Low Countries, where the recognition of the existence of the United Netherlands – in the Treaty of Munster

(1648) – opened up a half century of sporadic warfare for the mastery of the prosperous cities of north-west Europe.

The principal technical improvements were in the range and hitting power of artillery, and especially in the application of the appropriate applied mathematics by artillery specialists, with resulting dramatic improvements in accuracy. In addition, all this was achieved while improving the mobility of field and siege artillery pieces. A second series of changes, less clearly beneficial to the attacker, was the emergence of entirely musket-armed infantry.

The reaction in fortification design was the development of more and more complicated structures beyond the city walls, which endeavoured to keep the city outside the range of enemy artillery while presenting a series of obstacles of increasing strength that would need to be successively stormed before the walls themselves could be approached. Something of the resulting complexity, and its associated terminology, in this 'age of codification and formalism' leading to a 'babel of technical terms and bizarre geometry' (Brice 1984: 115) can be appreciated in Figure 2.10. However, such symmetry and completeness was rarely achieved outside the pages of the engineering instruction manuals – in this case drawn from Coehoorn's (1685) *Nieuwe Vestingbouw op een Natte of Lage Horisont* ('New Fortifications for Wet or Low-lying Areas'), which was especially appropriate in the Dutch environment. The walls, arrow-head bastions and even citadel of the previous century are now so surrounded by outworks as to have largely lost their function. Defending artillery is located in a series of attached ravelins, detached lunettes, multiple horn works, and crown works, defending the more vulnerable points, especially the entrances, reinforced by the smaller contre gardes and tenailles. The two locating principles were (1) that each face should be defended from assault by covering fire from a position behind it, and therefore presumably not yet under attack, and (2) that in the event of it being lost, each position could be easily swept by artillery sited above and behind it.

The idealized cross-section through a typical defence work is as complex and scientifically determined as the two-dimensional plan. The defenders present the lowest possible target to enemy fire, protected by the slopes of the glacis and plongee, while themselves exploiting the clear field of fire on an enemy assaulting across those slopes. The ditch was most usually water filled in The Netherlands (known as a gracht). The main distinction between Coehoorn and Vauban is that the former was generally 'wet' and the latter 'dry'. Water, both within the grachten and also as controlled inundations in suitable areas around the defences, provided extra obstacles, if the water levels could be managed and kept ice free, which was frequently difficult to achieve – see Koeman-Poel's (1982) account of the constant threat to the fortress of Bourtange offered by the freezing of the marshy defences around it.

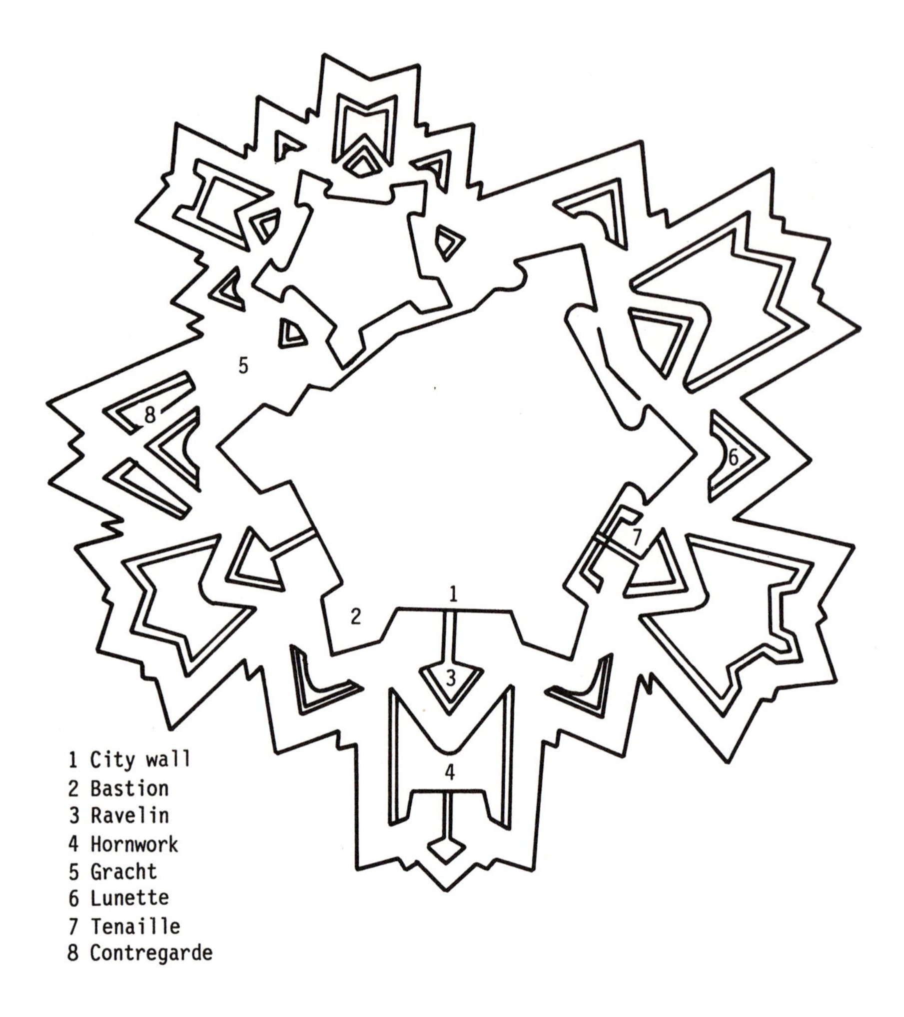

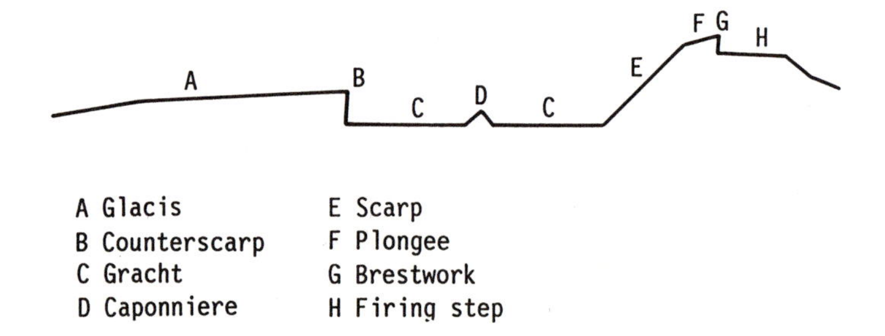

Figure 2.10 Idealized Coehoorn fortification system, 1685

Attacking infantry, having survived the killing ground of the glacis and traversed the ditch, were then fired on from behind, while engaged in climbing the scarp. The counter-scarp gallery provided covered positions for muskets firing inwards against the scarp; in dry ditches, caponnieres built across the ditch provided similar positions for sweeping the ditch itself. The musket was thus effectively exploited in static defence over distances well within its effective range, with protection offered during the lengthy reloading, while attacking infantry had few chances to retaliate.

If effectively manned, adequately provisioned, and not betrayed, such a fortified city was proof against all but the most determined and lengthy attack. Small wonder that 'under such conditions, war consisted primarily of an endless series of sieges' (Creveld 1977: 9). As Mellor (1987) has argued, cities acted as fortified magazines, and, as it was almost impossible for a field army to operate at more than about a week's march from them, then warfare inevitably became a static affair. Creveld (1977) has calculated that the eighteenth century witnessed some 60 battles in Europe but more than 200 sieges, the majority of which were unsuccessful.

The application of these theoretical designs to the defence of actual cities, however, was constrained by the variations of the particular sites (which modified the geometrical symmetry of the textbooks) and even more so by the enormous expense involved. The building effort involved determined that the complete system was only constructed in a very few carefully selected sites where major national defence interests were clearly threatened, and a large proportion of the national defence budget should be expended. The investment of money, advanced technology, and large quantities of land for the protective encirclement of a relatively small urban area makes it clear that the defence objectives of these works were national rather than local. The defence of the city itself was an incidental result of the creation of a fortress of value in a wider strategic context.

The modification of existing fortifications was a more usual and cheaper alternative to a complete rebuilding. In particular, many cities added some of the outworks favoured by the new system to their existing defences – incrementally, as finances, land availability, and external threats prompted. Utrecht, for example, had added hornworks by the middle of the seventeenth century and detached lunettes in the course of the eighteenth (Sneep *et al.* 1982).

The southern part of the Low Countries provides some of the best examples of purpose-built fortifications of this period – if only because of the critical strategic importance of this region to the countries which led in the field of fortification technology. French ambitions for expansion to 'les limites naturelles' threatened the survival of the Dutch Republic, and led to the zone of fortified cities from the Flemish coast at Nieuwpoort through Oudenaarde, Mons and Charleroi in Hainault, to the Meuse at Namur and Moselle at Luxembourg. These were the twelve 'barrier fortresses' whose possession was the object of 150 years of campaigning and which were garrisoned more or less continuously by the Dutch – from the Treaty of Rijswijk (1697) to their final abandonment as technically and strategically obsolete in 1781.

An important extension of this position – covering its eastern flank – was Maastricht, known as 'the key to the Meuse', because of its control over (1) the east–west routes between the Palatinate and Belgium and (2) the north–south links between the Walloon heartland of the Sambre-Meuse and the central Netherlands. The history of the fortification of this important position extends from the Roman-defended bridging point of *Trajectum ad Mosam* to the fortress of Eben Emael, whose seizure in May 1940 was the first necessity for the German invasion of Belgium. It also, however,

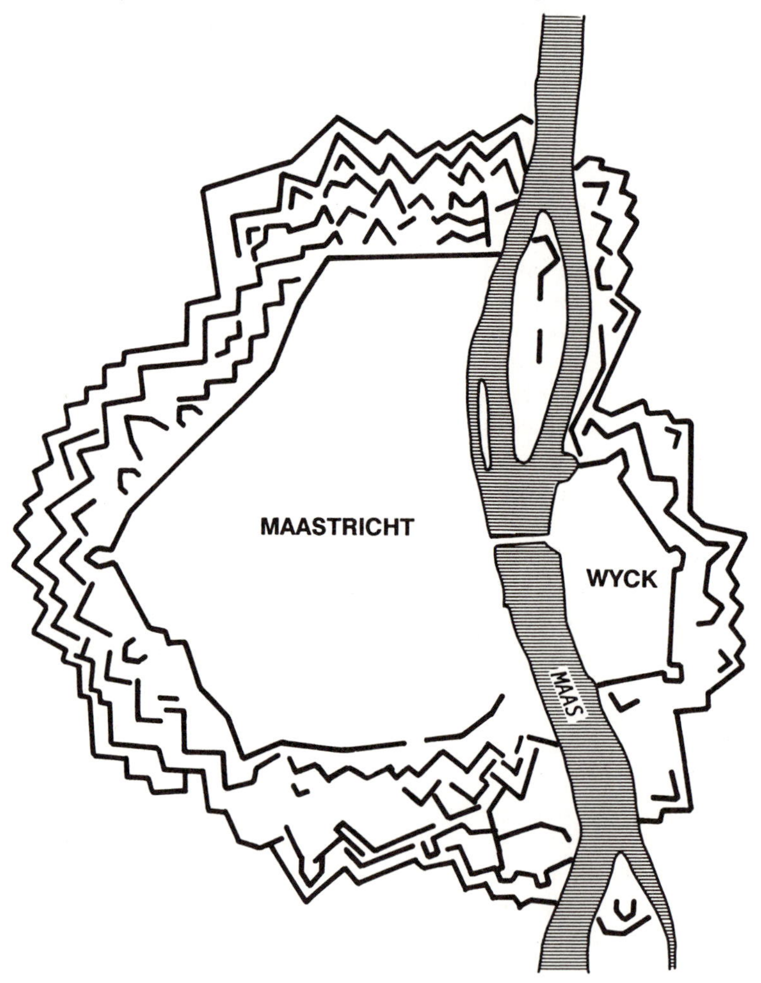

Figure 2.11 The fortifications of Maastricht, 1794

provides one of the most complete examples of the fortification system of the late seventeenth century. Incorporated into the Dutch republic by conquest in 1632, this southernmost extension of The Netherlands was successfully besieged by the French in 1673, and the fortifications completely rebuilt under Vauban's supervision in 1697. Figure 2.11 shows the result as it existed with few modifications in 1794. No fewer than seventy-six outworks of one sort or another were built. The river was not incorporated as part of the defence works, except to feed a number of grachten and inundate some of the surrounding land in emergency, but was enclosed with the suburb of Wijk within the defences. The plan shown in Figure 2.11 shows only one aspect of the fortification system (namely, that visible above the surface). There is an equally complex system of underground passages and chambers developed for two purposes: (1) as a means of communication between the outworks for the garrison and for manning the counter-scarp galleries; and (2) as part of anti-mining warfare. Underground passages extend under the outermost glacis to provide listening posts and launching points against enemy miners (Morreau 1979).

Few examples of this fortification system (in such a complete condition with the full range of possible defensive structures, taken directly from the most advanced contemporary texts on military engineering) were constructed for cities as large as Maastricht. In The Netherlands two much smaller fortresses were rebuilt by Coehoorn to cover the eastern approaches into Germany – namely, Coevorden and Bourtange (Zuydewijn, 1977; Koeman-Poel, 1982) – but cost alone determined that these would be rare occurrences.

More typical is the application of the system to a provincial capital, such as Groningen, which although the key to the defence of the northern Netherlands was also a flourishing regional commercial and manufacturing centre and part of the Hanseatic trading system. The difference in scale is clear between the medieval walls (last reconstructed in 1470, whose circuit is clearly visible in Figure 2.12) and the simple but enormous 'new' wall with its seventeen bastions and broad surrounding gracht, constructed by the engineer Geelkerk between 1616 and 1625. Prudently, the area of the city within the walls was doubled by northern and eastern extensions which provided a land reserve still not exhausted 200 years later. Most of the technical embellishments noted in Maastricht are missing in this provincial application but provision was included for the inundation of land outside the walls – except on the southern side, where the land rises significantly. This last weakness was a long-standing concern of the city administrators. In 1688 an attempt was made to dig out and remove the Kempkensberg ridge, in order to prevent the siting of enemy artillery and allow inundation. Finally, in 1700, an appeal was made to Coehoorn himself, and a new separate defence line 2 kilometres south of the city (the *Helpmanlinie*) was built to block the southern approaches (see Schuitema Meijer 1974). The city has been besieged three times, and on each occasion was attacked from the

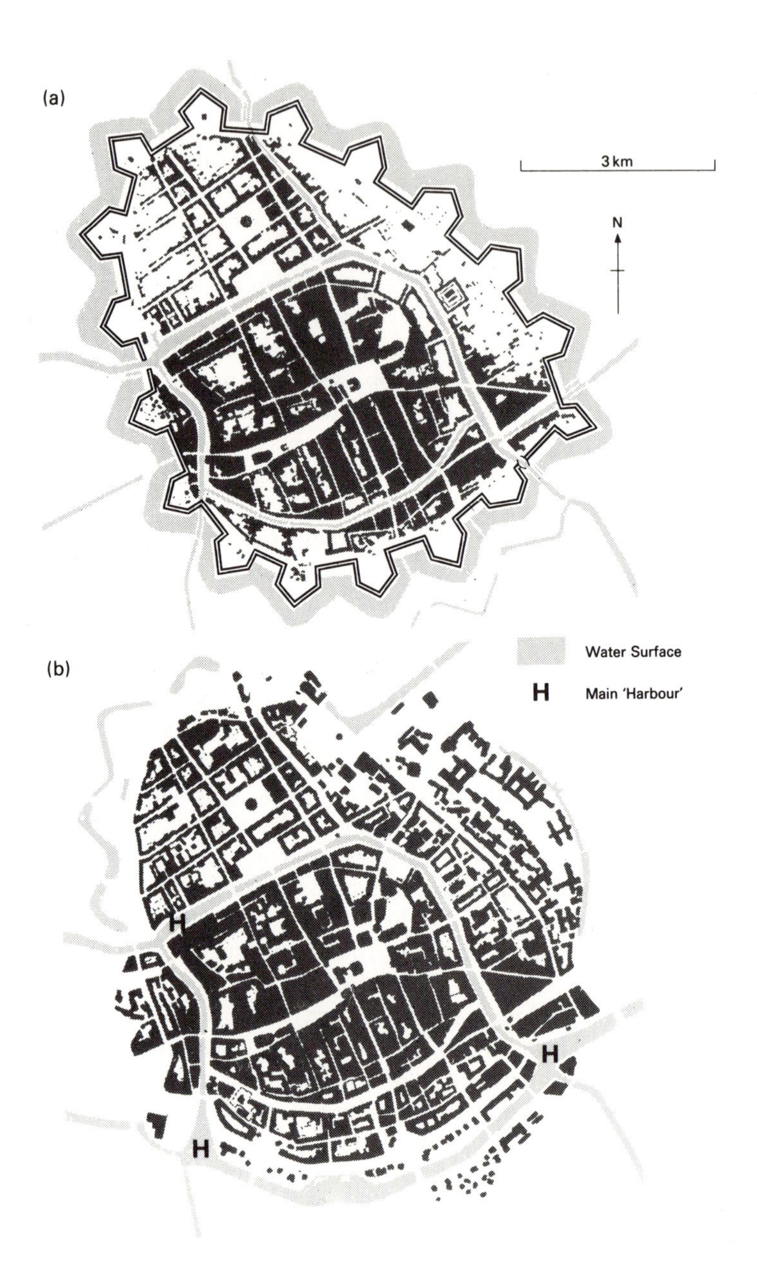

Source: Smook (1984)

Figure 2.12 Groningen (The Netherlands): (a) fortifications in 1826; (b) use of fortifications, 1980

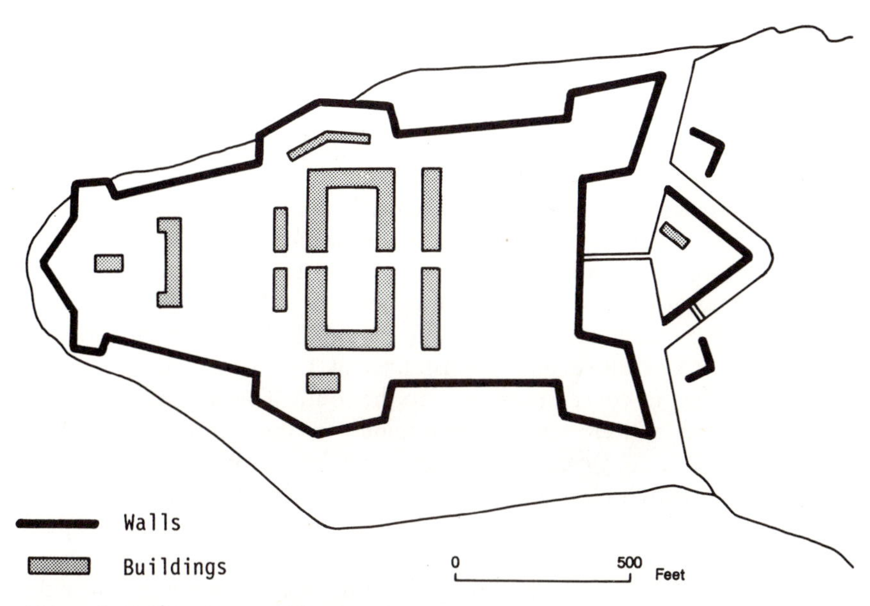

Figure 2.13 Fort George (Scotland)

southern side: in 1594, as a Habsburg city, it fell to the army of the States General; in 1672 it successfully resisted a German army; and in 1945 a German army capitulated to a Canadian assault (see Chapter 5).

The only notable British example from the period is exceptional in many respects. The need for fortified towns had, in general, long since passed in Britain when the Jacobite disturbances of the first half of the eighteenth century led to a determination to 'pacify' the Scottish highland clans by military occupation. To this end a system of roads and purpose-build garrison towns for an army of occupation was constructed. The clearest surviving example is Fort George (Figure 2.13), which was built in 1748 and incorporated many of the features of the continental fortification system of a generation earlier (Kightly 1980). The promontory site was occupied by a walled settlement with four bastions, two demi-bastions and a point battery covering the seaward approach. The only land approach was covered by a full range of outworks, including ravelin, lunettes, glacis and ditch, swept by counter-scarp and traverse fire. That it was felt that the 1,600 soldiers and 80 cannon needed such protection against the wild but lightly armed Scottish highlanders was a reflection of official panic rather than any real possibility of attack by a technologically advanced enemy.

The rivalry between Britain and France for the domination of North America, which reached its decisive climax in the first half of the eighteenth century, resulted in the export of European urban defence techniques to the New World. The most architecturally impressive result of this is Louisbourg on Cape Breton Island, which has been called 'the only fortress town ever built in North America' (Downey 1965), a claim justified by its simple purpose of being the impregnable corner-stone for the defence of New France. Begun in 1719, its location was intended (1) to control the entry to the St Lawrence routeway into the heart of the French position in North America, and (2) by its existence, threaten the security of the British settlements to the south, in mainland Nova Scotia and even New England. This threat was taken so seriously that it was attacked and overrun by New England militia while still in process of its painstaking and, owing to the marshy nature of the site, difficult construction in 1745, but it was handed back to French control in 1748. The resulting dismay among British Americans led in turn to the establishment in 1749 of a British counter-fortress, Halifax, to cover Louisbourg to the south. The layout of the town and its defences as it appeared in the middle of the century is shown in Figure 2.14. While not particularly impressive by European standards, the fortification system is clearly derived from the earlier works of Vauban. The landward side of the peninsula is covered with the full sequence of rampart, bastion, ditch, counter-scarp and glacis, while the seaward side also has a detached ravelin protecting the gate. Within the walls, a town to house the 6,000 inhabitants of the military garrison and civil administration was laid out on a 57-acre site to a rectilinear pattern of streets focusing upon the citadel of the King's Bastion and the parade ground below it. Only in the northern part of the town near the harbour, where a commercial fishing settlement developed, does the military function, and with it the regular street pattern, become less pronounced. Given the importance of Louisbourg in the eyes of both the French (who made such a concentrated investment of resources) and the British (whose North American strategy was dominated for 40 years by the threat of its existence), it is surprising that it fell to Wolfe in 1758, in an attack lasting only 8 days. In part, this was confirmation of the truism that a fortification is only as sound as the men who man it and the commander who organizes the defence – and Louisbourg was badly served in both respects. Equally, however, it was structurally less impressive than contemporaries claimed, and (by European standards) obsolete in a number of respects. It was dependent upon supply by sea, control of which was lost, and the harbour to the north was inadequately covered by artillery, so that the defending fleet was easily disposed of. Once artillery (including mortars) had been placed on the landward side, the absence of detached outworks allowed the town to be brought well within bombardment range (Downey 1965).

The fall of Louisbourg, and its immediate dismantlement, signalled the collapse of the French position in North America. Quebec City fell in 1759 (significantly to a battle outside the equally obsolescent fortifications). Montreal, with only rudimentary fortifications, offered no effective resistance in 1760. However, this was not the end of the history of urban fortification

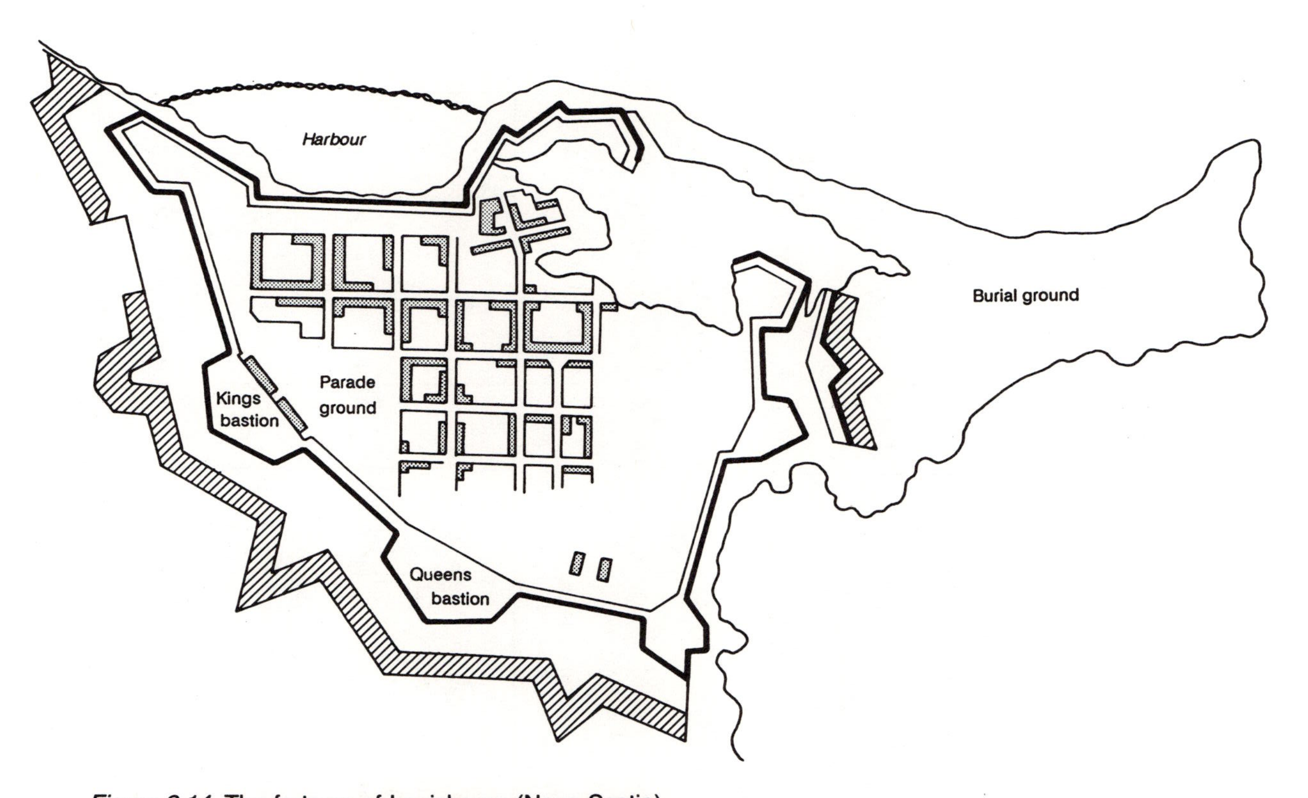

Figure 2.14 The fortress of Louisbourg (Nova Scotia)

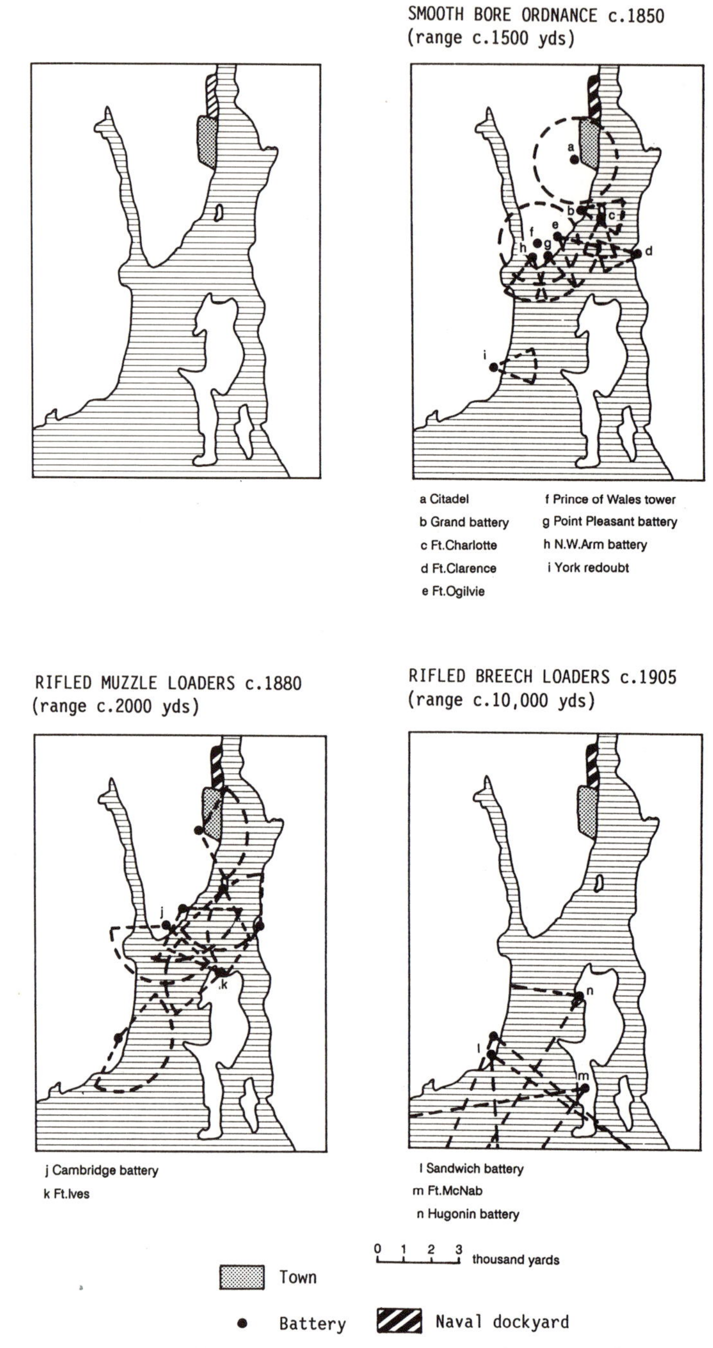

Figure 2.15 Evolution of the artillery defences of Halifax (Nova Scotia)

in the region, as Anglo-French rivalry was replaced by Anglo-American. The present surviving walls of Quebec City were British-built between 1823 and 1832 as a defence against the very real threat of American invasion, as were those of a number of settlements on the Niagara peninsula and the northern shore of Lake Ontario – including Kingston, and York, later Toronto (see Kerr 1960 and Nader 1976). The most enduring example, however, of a fortified town in British North America is Halifax and its associated naval base and water approaches (see Figure 2.15). As one corner of the Halifax–Bermuda–Gibraltar triangle of strategic bases for the control of the North Atlantic, it was the principal garrison of British North America (from its foundation in 1749 to Confederation in 1867), with its defences being continuously modernized throughout this period. The citadel (rebuilt a number of times between 1749 and 1828) dominates the town and harbour below, while entrance to the harbour is covered by a number of self-contained artillery positions of different periods, including the York redoubt of 1793 and a Martello tower of 1796 (Piers 1947).

The last fortified cities

The pace of change in urban fortifications through the nineteenth century and beyond was largely dictated by important changes in artillery technology. Until around the middle of the century such changes could be seen broadly as an evolutionary continuation of the trends already apparent for more than 100 years. Indeed, the cannon used in the Crimean War (1854) were smooth bore, cast-iron ordnance that would have been not notably unfamiliar to a gunner from the Spanish Armada (1588).

Soon after the middle of the nineteenth century, however, a number of more-or-less simultaneous and related developments quickened the whole tempo of development, creating an entirely new situation. These were the rifling of gun barrels, the use of steel, breech loading and the exploding shell. Together, these increased the hitting power of artillery over much longer ranges, and the average weight of projectiles increased eightfold during the second half of the century. In 1850 the effective range of heavy artillery was still only 2–3 kilometres, fractionally more than it had been 50 years earlier, while 50 years later it had increased fivefold. A race developed between the efficacy of artillery and the counter-improvements in fortifications to withstand it; a race that ultimately the latter could not win as stronger, and more expensive, works were required, at ever shorter intervals of time, only to be rendered obsolete before they could be completed.

There was an obvious relationship between the rifling of barrels, the use of exploding cylindrical shells rather than circular shot, and loading through a movable breech rather than down the muzzle. Shells could be rather awkwardly screwed down a grooved barrel but the shell using detachable propellant in a separate casing clearly required a breech for discharging them. However, rifled muzzle-loaders were installed in a number of fortifications, notably to defend the naval base of Halifax, although breech-loaders were available (Johnston 1981). The choice was influenced by more than military conservatism, or colonial backwardness. It was also influenced by the need to ensure reliability and rate of fire: both being critical in the defence of naval bases against closing targets. The reliability and accuracy of early breech-loaders were suspect. Furthermore, the technical development implied an equally important revolution in the way they were served – the speed of adoption being often constrained as much by the availability of trained crews as by the equipment. However, by the 1880s all three developments had been perfected, integrated and applied – leading to dramatic improvements in range, accuracy, hitting power and, ultimately, rate of fire (Allen 1976).

The effectiveness of these improvements in artillery equipment was, to a large extent, dependent upon a wide range of other developments, such as accurate maps and electronic communications. As the increases in range took targets beyond the vision of the gunners, fire-control systems became more important.

To these revolutionary developments in artillery can be added two other important military inventions of the last quarter of the nineteenth century. First, there was the development of a rapid-fire, anti-personnel weapon (the machine gun) and, second, the military use of barbed wire. Taken together, these developments effectively neutralized the open space between the fortified positions, rendering infiltration by unprotected infantry extremely costly.

The tilting of the balance between defenders or attackers as a result of all these changes depended largely on whether they could be applied more effectively to fixed-fortress or to mobile-field artillery. (In the second category the use of large-calibre guns on armoured warships was a particular concern of coastal defences.) The result was an arms race whose beginnings can be detected from the middle of the eighteenth century, when the great fortress towns still dominated European warfare, and whose outcome ultimately rendered urban fortifications impossible. The recognition that artillery was the critical weapon (and infantry merely an exploiter of success) determined that the bastion evolved from being a means of defending the walls from escalade to the point of defence or attack itself. It evolved as a self-contained artillery battery located ever further from the walls themselves in response to increases in the potential range of bombardment. The question was thus raised for the first time in many thousands of years as to why city walls were needed at all.

If the defence of the city was now in the hands of its artillery batteries located well outside the city, successful attack depended upon the silencing of those batteries, and not on the assault of infantry upon the walls through breaches created by artillery. Thus there developed a somewhat paradoxical situation in Europe during the second half of the nineteenth century: on the one hand, increasing attention was paid to the defence of the fortress cities, while, on the other, there was a widespread movement for the abandonment and dismantling of city walls.

What became known as the 'Prussian System' was a ring of detached forts spaced around the city, covering with their artillery the open territory between and in front of them. The distances between such forts, and between the defensive ring and the city itself, was a function of the range of gunfire. In 1800 some 600–900 metres was considered ideal but by the 1830s, when the defences of Antwerp were reconstructed, this had increased to around 1,400 metres. Developments after 1850 made such distances (and the forts covering them) obsolete. By 1865 a new ring of nine forts around Antwerp, each mounting more than 100 guns, was constructed at 3–4 kilometres from the city; by the 1880s the defences of Liege and Namur were planned at 4–9 kilometres; and the final rebuilding of Antwerp, completed in 1904, included sixteen forts at distances of 10–15 kilometres (see Figure 2.16). As such distances increased with each refortification – involving ever more extensive areas of land as well as investment of national capital – it is obvious that fewer cities could receive such attention. Such defences therefore become so costly and widespread as to merge with

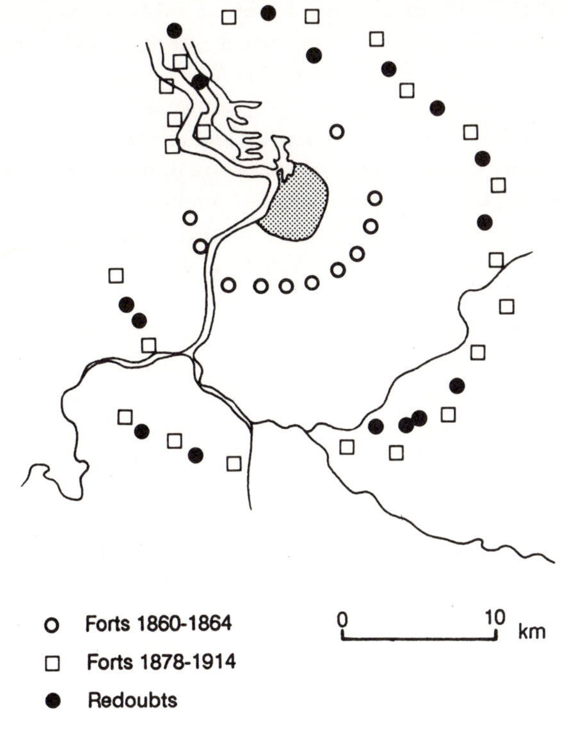

Figure 2.16 The last fortifications of Antwerp

regional and national defence lines, so that it becomes less realistic to think in terms of the fortification of a particular city as such but rather of defence systems designed to protect urban regions.

It is thus inevitable that examples become fewer and necessarily rarer. Naval bases have always had a particular need for secure defences without which the fleet cannot operate, and it is therefore not surprising that Portsmouth harbour, a fleet base since Roman times, should have received constant attention in this respect through the centuries (see Chapter 3). The British mid-Victorian world empire depended upon the Royal Navy which in turn depended upon the safety of its bases, of which Portsmouth was the most important and best protected. The system was essentially a complete ring of artillery forts. These covered the landward approaches to Portsea Island over the crest of Portsdown Hill (the seizure of which would have allowed the dockyard to be fired upon); the landward approaches to the Gosport peninsula and western shore of the harbour; and the seaward approaches (through either Spithead or the Solent) which involved both land-based batteries and artillery forts in the water covering both straits. This most complete defensive system of the period was initiated in the 1860s, made effectively obsolete by developments in naval gunnery almost on completion, and was never tested by action (Corney 1983).

The sieges that did occur during this period, however, served to underline to contemporaries the continued importance of static defences, if properly manned. For example, the Crimean War (1853–6) resolved itself into the long siege of the naval base of Sebastopol, whose defences were by no means as advanced as the 'Portsmouth System'. The effectiveness against superior forces of even quite minimal, hastily erected, earth defences was demonstrated in the American Civil War at the siege of Vicksburg on the Mississippi in 1863 and at the less well-known, but strategically as important, siege of Petersburg (1864–5), covering the southern approach to Richmond. The former held for 45 days on a fortified line only 2–3 kilometres from the town centre, and the latter involved 10 months of siege behind a network of some 80 kilometres of trenches and earth redoubts (Mitchell 1955). Even 50 years later, the Russo-Japanese War was dominated in 1904–5 by the 11-month-long siege of Port Arthur, the principal Russian Pacific naval base, which had been specifically fortified for this purpose and whose possession was essential for the success of the Japanese invasion.

The 130 years from the end of the Napoleonic Wars to the end of the Second World War was a period of transformation. On the one hand, city walls (having largely lost their military functions) were in the process of being dismantled, while, on the other, confidence was placed in static defences around cities as part of national defence policies. The campaigns of Napoleon may have marked the 'end of the era of siege warfare' (Creveld 1977: 40), as far as the walled city was concerned, but the next century witnessed the most extensive, and expensive, construction of fixed fortifications in history.

Fortified cities were incorporated into a broader idea of the national redoubt, of which some of the best examples are in north-west Europe, where military architecture was seen as an alternative to manpower. The attempt of France to defend its eastern frontier is perhaps the best known example, where the existing fortress towns of Alsace and Lorraine were incorporated during the inter-war period into the Maginot Line, which has given its name as an archetype to a whole category of defence policies. For countries such as Denmark, Belgium and The Netherlands, the step of linking together the expanding defence systems of cities (which had long been fortified to enclose whole urban regions) was particularly attractive. In each case such a reliance on fixed, prepared, positional defence was a reaction of small countries to a larger and potentially aggressive neighbour. In all three cases, although at different times, the neighbour was Germany.

As far as Denmark was concerned, the idea of fortifying the narrow base of the Jutland peninsula (the only landbridge with the continent) is almost as old as the state itself. The *Dannevirke*, a defensive line demarcated by a series of ditches and embankments, was built originally between AD 810 and AD 1160. In the early nineteenth century a relatively small and weak Denmark, in the shadow of an emergent Germany, reactivated the old defence line with a zone of fortress towns in depth – such as Slesvig, Dybbol, Fredericia and Sonderberg – extending from the Baltic to the North Sea (Skovmand 1980).

The Dutch case demonstrates the most complete reliance on such a single-minded, national-defence policy throughout this period. It could be argued that this attitude was by no means new but was a return to the seventeenth-century 'barrier fortresses' – although, as the nineteenth century proceeded, the defence was orientated to protect the eastern, rather than the southern, approaches, as Germany, rather than France, appeared the more likely aggressor. Such a national system, officially named *Vesting Holland* (Fortress Holland) in 1874, could be seen as only a larger version of the skilful exploitation of water barriers that previously had protected single cities scaled up to defending the close cluster of important towns in the western Netherlands.

Figure 2.17 shows the main defence lines in this system which were in operation from 1815 to 1940, although with varying emphasis and frequent rebuilding during this long period (Rolf and Saai 1986; Buist 1987). The basic defensive principle was simple. A water barrier (deep enough to obstruct passage and wider than the current range of field artillery) had to be capable of being created quickly in emergency by controlled inundation, mainly along the lines of the existing Maas and Rhine distributaries. The sluices (which regulated this water management) and the few causeway routes through the barrier were then strongly fortified. Although the architectural details of these strong-points, and their locations, varied through the many rebuilding periods, this system, conceived on the basis of

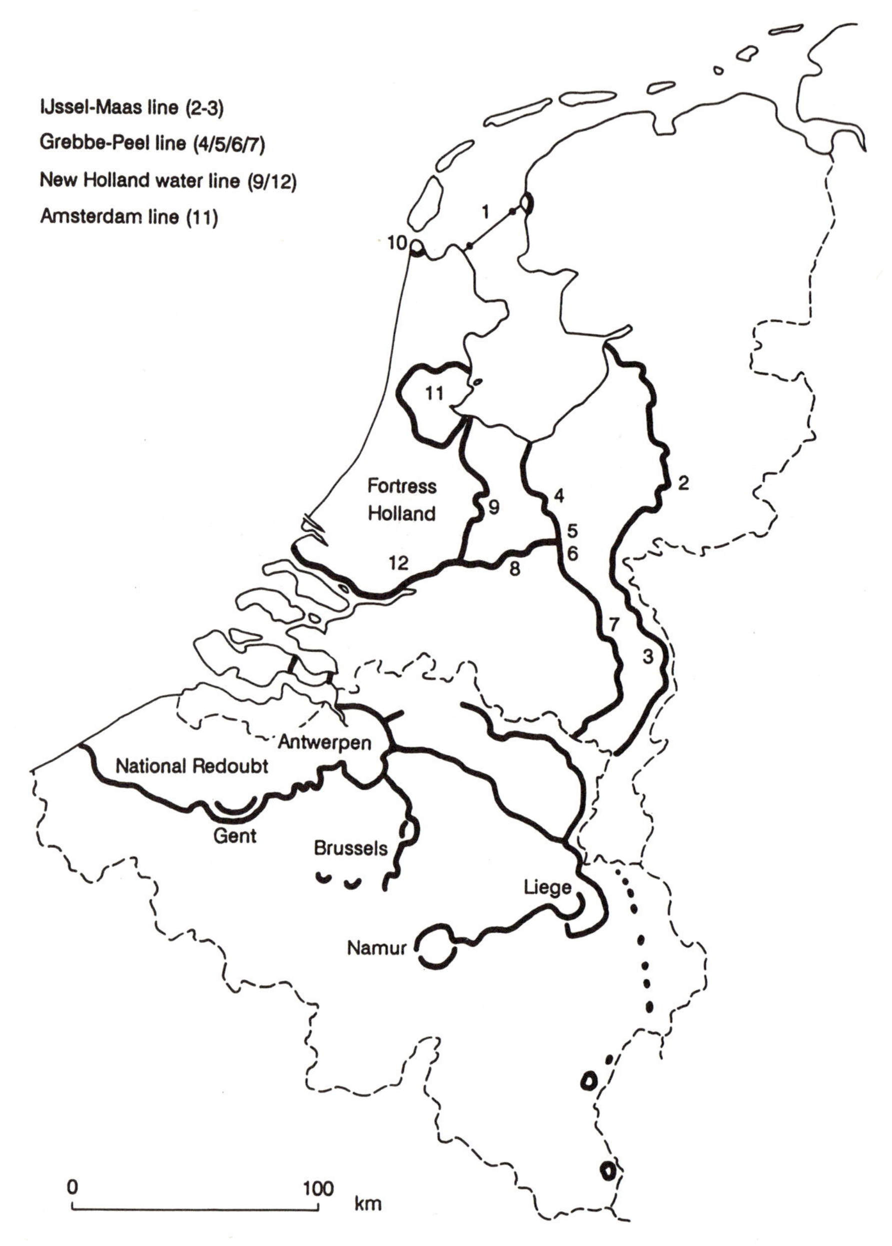

Figure 2.17 The national defence lines in The Netherlands and Belgium

Napoleonic experience was essentially that which was defended against the Wehrmacht in 1940.

The easternmost *IJssel-Maas* line was essentially a delaying position, to allow time for the inundation process to be completed further west and its defenders mobilized. The towns controlling the crossing points along these rivers were specifically excluded from the dismantling legislation that allowed other cities in The Netherlands to remove their walls in 1874. The western Netherlands was protected on three sides by major water surfaces: the North Sea; the Zuiderzee; and the Rhine. The most vulnerable stretch, therefore, was the 30 kilometres between the Rhine and the southern shore of the Zuiderzee. Here the *Nieuwe Hollandse Waterlinie* was built in 1815 – and improved in 1824, 1840, 1860 and 1880 – with its southernmost point anchored by the incorporation of the ring of forts defending Utrecht. By the 1930s, a second line, the *Grebbelinie*, was built to the east of the *Waterlinie* as a reinforcement, and ultimately replacement, to it. The building of the enclosing dam across the Zuiderzee in 1936 had created a new point of entry into Fortress Holland which was protected by new defences at Kornwederzand. The naval base at Den Helder and the capital, Amsterdam, received separate attention as inner redoubts within the national fortress. The final renovation of Amsterdam's defences was in the period 1883–1914, when no less than thirty-seven artillery forts were located at distances from 10–15 kilometres around the city. The fortress of Amsterdam was not formally decommissioned and its defences finally abandoned until 1922.

The Belgian national redoubt was similar to the Dutch in conception, motivation and orientation eastwards but depended much less upon inundation (Rolf and Saai 1986). It consisted of two main lines (Figure 2.17): (1) the delaying position of the Meuse which linked the key fortresses of Liege and Namur, and was later extended north-westwards along the Albert Canal; and (2) an inner redoubt based on the formidable defences of Antwerp and Ghent, linked by the river Lys, with its southern flank protected in part by inundation. The timing of the construction and reconstruction was very similar to that of the Dutch, although there was no co-ordination between the two, with the last period being from 1920 to 1936, when a third position, the KW Line, running from Koningshoyct to Wavre, was added.

The argument over the effectiveness of such defences continued throughout the period, as well as after they were tested in action. In the Danish case this was between 1849–50 and 1864, in the Belgian and French cases in 1914 and 1940, respectively, and in The Netherlands in 1940. The Danish *Dannevirke* line (or, more properly, defensive zone) was successfully held against superior German forces in 1848–50 but succumbed to the Prussians in 1864. The Belgian Meuse forts around Liege and Namur were creditably defended, as was Antwerp, in 1914. Even in 1940 the Namur forts took 13 days to clear, while one of the Liege forts was still in Belgian hands when the king capitulated the army (see Belgian Ministry of Foreign Affairs 1941). An ill-prepared, obsoletely armed and poorly led Dutch army was surrendered in 5 days in 1940, but not because Fortress Holland had been successfully overrun, and where it was assaulted at Kornwederzand and on the Grebbelinie, it was determinedly defended. Rather, it had been overflown, by both air-landed troops and by bombers, and the western cities held as hostage to aerial attack (Kleffens 1940; de Jong 1970). In the French case, the existence of such defences largely dictated strategies of attack to avoid them, and in 1916 Verdun demonstrated the cost of attacking a determinedly defended fort system. In all these cases the principal cost of the reliance on prepared defence lines was that they pinned down available defenders, while leaving mobility and flexibility to the invaders.

It is clear that this long chronology has now reached the point where the range of artillery, the mobility and sheer size of manageable forces engaged, has so enlarged the whole scale of warfare (and with it the scale of fortification) as to make it difficult to think in terms of the fortified city. A succession of short logical steps has led from the walled city, through the detached bastion and the defending fortress ring to the Waterlinie, the national redoubt, and the Maginot line, and then to the continental scale defences of *Festung Europa* behind its *Atlantic Wall* in 1944. The spatial patterns created by each of the phases has been generalized in the model in Figure 2.18. Such steps, however, have now led well beyond the defence of cities as such and thus beyond the scope of this book.

EFFECTS OF FORTIFICATIONS ON THE CITY

The existence of fortifications had a number of important consequences for the development, functioning, structure and form of the city so defended, some of which were intentional and a part of the defensive scheme itself, while others were frequently unforeseen and unintended by-products.

There were a large number of indirect effects, the most important of which was the fundamental question of finance. Defence works were a capital investment whose cost, always a large percentage of local incomes, escalated sharply as military architecture responded to the threat of artillery. An important question was always, who pays? Although most medieval towns were responsible for the financing of the construction of their own defences, the complex masterpieces of the seventeenth and eighteenth centuries, described above, were well beyond the financial resources of all but the richest city-states and were usually national projects, frequently consuming large portions of the national income. In either event fortifications had to be paid for, whether locally or nationally, and formed a substantial tax burden that the trading cities were expected to shoulder. The initial costs were supplemented by almost continuous maintenance costs which were more likely to be a local responsibility even when the original investment was national. The case of Utrecht described earlier, where

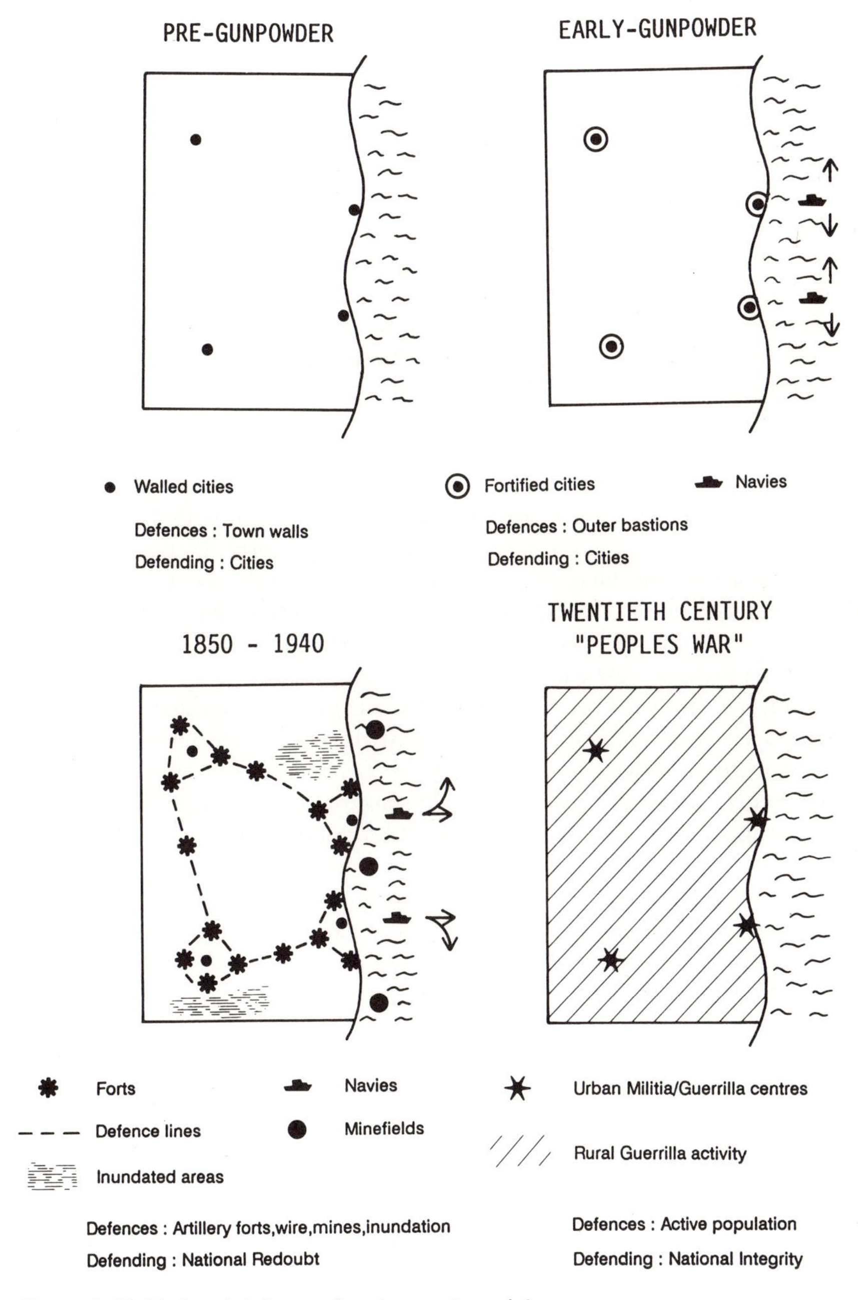

Figure 2.18 National defence development model

maintenance was delegated to the merchant guilds, each responsible for a measured section of wall and towers, still resulted in regular complaints in the city council about crumbling walls, infilling water defences and the escalating costs of repairs (de Vlerk 1983).

A theme running through the history of fortifications has been the obvious truth that defences are only as good as the defenders who man them and many of the most 'impregnable' fortresses described earlier fell with nonchalant ease to determined assaults when they were inadequately manned. The fortress town of Naarden, for example, is frequently presented as an illustration of perfection in military engineering. Its historical record, however, shows scant return on the investment of skills and capital in it. Its first complete defence system of curtain walls, towers and gracht built in 1560 was surrendered without demur to the Spanish in 1572. Although redesigned by Anthonisz in 1574 and modernized to the highest contemporary standards in 1668, its garrison was successfully bluffed into evacuation in 1672. Finally redesigned by Coehoorn in 1674–85, it was entered unopposed by the French in 1794 (Vrankrijker 1965). The general problem was that, throughout almost all the period discussed, the trained personnel available – whether drawn from a knightly class fulfilling its feudal obligations, a limited cadre of full-time professional soldiers or hired mercenaries – were used to form the mobile field-armies: they could rarely be spared for static garrison duties, for which they were in any event unsuited by training, inclination and sheer cost. The settlement of discharged veterans on land grants in exchange for garrison duty was one solution to this problem used by empires from the Roman to the British for the provision of a trained manpower reserve along vulnerable frontiers (see Chapter 3). But this was by its nature a wasting asset and most local defences, in most periods, were expected to be locally manned by city militias, trained bands and the like. These all encountered the obvious difficulties of conscripting and training an amateur garrison which in peace-time would be reluctant to train and in war-time equally reluctant to fight. Constantinople was reputedly the most strongly protected city in Christendom (protected by the famous triple land-walls of Theodosius and the less impressive single walls along the water defences of the Goldern Horn, the Bosphorous and Sea of Marmara), yet when its existence was threatened in 1453 it could raise only 4,973 willing citizens to man these defences out of a population of at least 100,000 (Fuller 1970: 363).

The problem increased steadily throughout the early modern period as a result of changes in society and military technology. The struggles of the cities for freedom from onerous obligations clashed with the increasing manning requirements of the ever more elaborate defence works. By the middle of the nineteenth century the complex system of outworks and forts demanded manpower beyond that which the city they were defending could possibly provide. The 1860 Antwerp fort system required a garrison of

70,000 (Brice 1984: 146). The even more complex Portsmouth fortifications would have needed around 100,000 men as a minimum garrison, if all positions were to be manned simultaneously. Such numbers could never have been provided by either the peacetime professional army or the local urban militia. One of the most important costs of the construction of the twentieth-century national redoubts of Belgium and The Netherlands, as well as the Maginot line in France, was that when put to the test of invasion they absorbed a very high proportion of the total national defence manpower, albeit a manpower whose lack of training and equipment often fitted it for little else. For example, of the twelve Dutch divisions mobilized in May 1940, eleven were committed to the various defence lines. This in turn, especially in the Benelux cases, resulted in insufficiently large field armies, which presented the enemy with the strategic initiative.

It was not only that defences became costly in terms of the quantity of manpower but also in terms of the skills needed. In the medieval city the skills, in terms of weapon handling, required of both attacker and defender were limited and approximately the same. The post-artillery fortifications, however, principally required skilled gunners rather than semi-trained citizen militias. The solutions were either, the diversion of a large part of the city's labour force into continuous military training, or, the maintenance of large standing professional garrisons for which the city would pay through taxation and more directly through the costs of billeting. Neither solution was attractive, which accounts for the almost continual friction between military and civil, and local and national authorities on this matter, and equally accounts for the seemingly inexplicable fall of many architecturally formidable defences to attack.

The most obvious, direct and pervasive effect of fortifications is the negative restrictions they imposed on the growth and development of the town. City defences are expensive fixed investments that cannot respond easily to the expansion, or conceivably also the contraction, of the urban built-up area. There are, of course, many examples of towns responding to pressures for expansion by the demolition of existing walls and their rebuilding to enclose a wider city. In fact, successive refortification through the centuries is more often the rule than the exception in the cases discussed earlier in this chapter, and pressures for urban expansion were frequently as important a motive for rebuilding as the changes in defence technology discussed above. Such refortification, however, was both expensive and troublesome (in terms of property rights), and such changes therefore occurred at infrequent intervals, separated by long periods of architectural stability. Fortification thus imposed a ratchet-like physical expansion on cities, punctuated by long periods when growth (with greater or lesser success) was contained within the walls.

Expansion beyond the walls was often discouraged by both the military and civil authorities, not only because it made the defence of such extra-mural districts more difficult, but also because it raised military and administrative problems. The military had an interest in maintaining uninterrupted fields of fire and denying cover, shelter and usable siege materials to an approaching enemy, while the civil authorities feared the weakening of their administrative and taxation powers over citizens living outside the walls.

The obvious result of an increasing demand for space, together with the discouragement of expansion beyond the walls, was a condition of almost continuous friction between those wishing to expand and those attempting to contain such expansion; the morphological balance resulting from these countervailing trends being largely determined by the strength of the urban growth, the effectiveness of the administration, and the state of national and local security. Times of peace and plenty led to suburban expansion, usually along the main access roads beyond the city gates, while the threat of war resulted in contraction and, if time permitted, demolition. It is to be expected that fortified cities would have a higher density of population and land use than those unfortified. Vance (1977) has argued that this difference

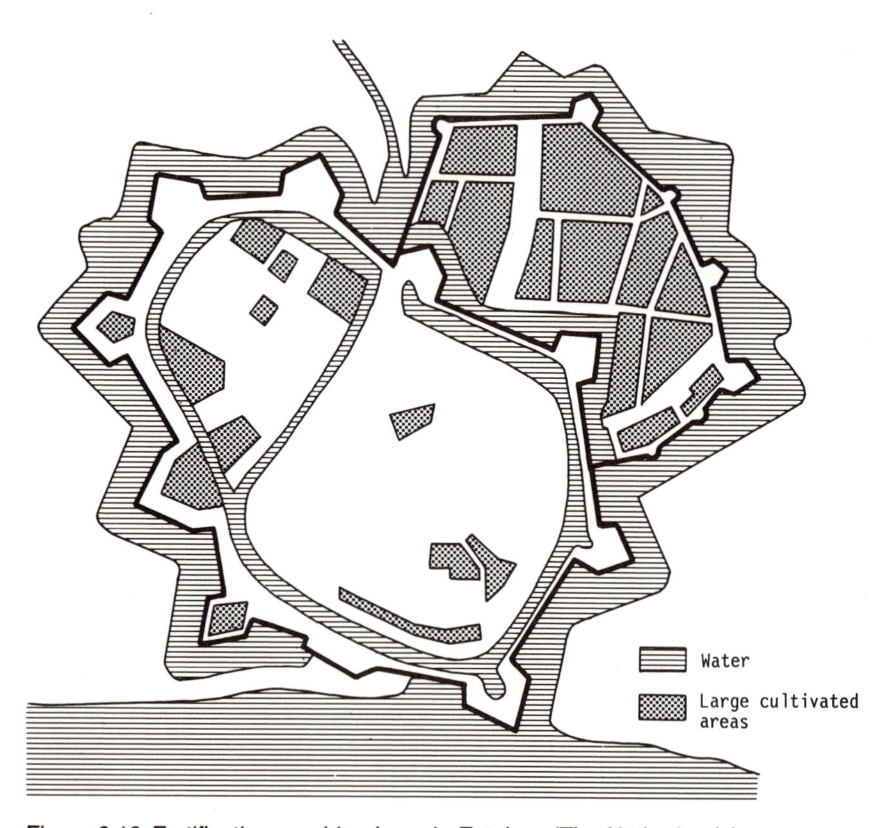

Figure 2.19 Fortifications and land use in Zutphen (The Netherlands)

in building density is a fundamental distinction between British towns and those of the European continent where fortifications survived much longer.

There are, however, many instances where planned expansion took the form of fortified suburbs rather than uncontrolled and undefended informal growth. For example, there is the city of Zutphen (see Figure 2.19), where in the thirteenth century the Dukes of Gelderland constructed the 'new town' (*Nieuwe Stad*) as a fortified extension to the existing town, to accommodate flourishing trading activities. By the end of the fourteenth century, however, this was insufficient, and a further extension (the *Spitaalstad*) was added (see Noordegraaf 1985). Similarly, the seventeenth-century contiguous 'new town' of Groningen is clearly visible in Figure 2.12 as a north-western extension, doubling the area of the town, when the new fortifications were built in 1616. The alternative to this incremental addition to the existing walls was a complete new circumvallation, allowing expansion in all directions (what might be termed the *onion*, rather than the *wart* technique). The best known example is probably Paris (see Bastier 1984) with its sequence of walls (see Figure 2.20). But Moscow is equally dramatic in its five successive concentric rings, beginning with the Kremlin (first masonry wall, enclosing approximately 30 hectares, in 1367) and ending with the 1520 walls enclosing a city of around 100,000 inhabitants and an area that sufficed until well into the nineteenth century (Morris 1972).

On occasion, the situation could be reversed, and an overambitious circumvallation was followed by a smaller than expected urban growth, or even a reduction in urban land-use demands, leaving the city to 'rattle around within its overlarge walls' (Vance 1977: 128). For example, Norwich not only remained quite contentedly within its twelfth-century walls through its fifteenth-century period of economic and population growth but had land to spare for most urban needs until well into the nineteenth century. The substantial areas of non-built-up land to be seen on many maps of walled cities may support the argument that the pressures for expansion were less than has often been argued. Equally, it must be remembered that many agricultural (and even recreational) activities were an essential urban land use (even in times of peace) and not necessarily a symptom of the existence of a land reserve within the walls.

It was not only the existence of the fortifications as such which imposed restrictions upon the expansion of the city but also the need to maintain a defensive zone around the city. Such a zone might include only the immediately adjacent space, or extend many kilometres from the city. The 500-year-long interaction between the offensive and defensive use of artillery (described earlier) led to the need for ever longer clear fields of observation and of fire. Carter (1983: 115) has described the seventeenth-century city as 'a small kernel within a very thick shell', and the thickness of the shell included not only the width of the fortifications and the length of their glacis approaches but also a further zone of restrictions – on the height and size of

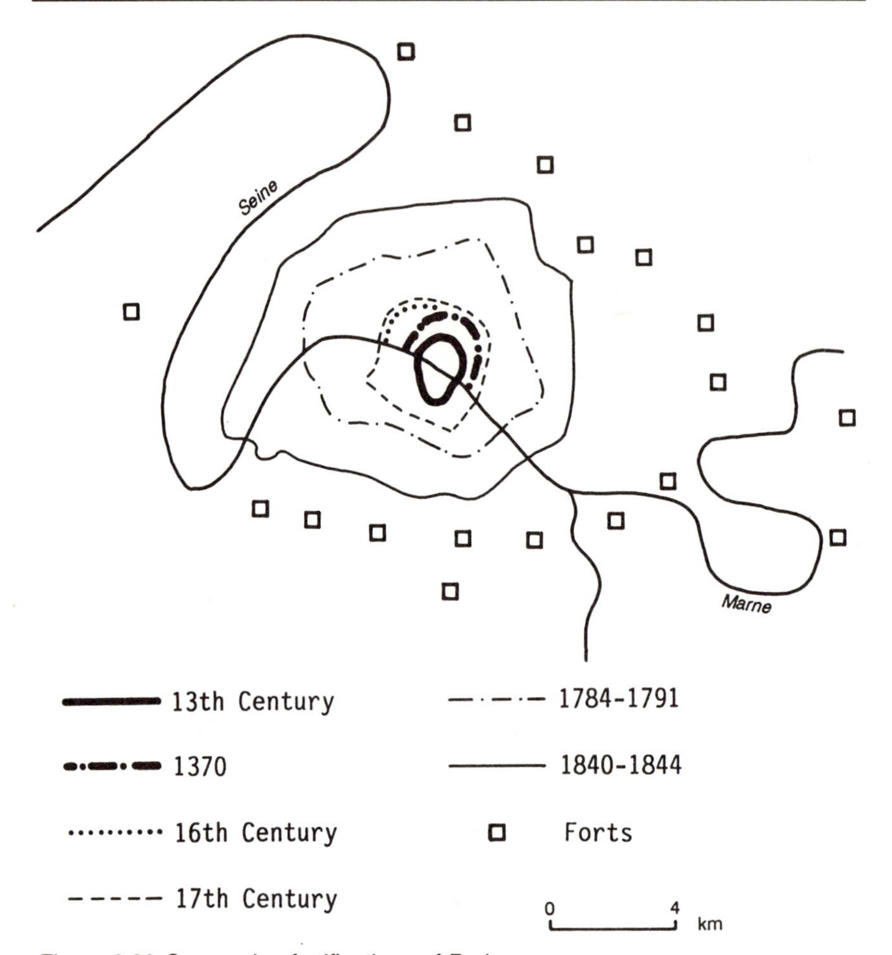

Figure 2.20 Successive fortifications of Paris

buildings, and even on the sort and permanence of the vegetation cover. A Dutch law of 1853, for example, specified three such zones within which restrictions of decreasing severity would apply, at 300, 600 and 1,000 metres, but in all of which only wooden buildings, quickly demolished in emergency, were permitted (Brand and Brand 1986).

The coming of the railway to towns which still possessed substantial fortifications often brought this problem to a head. One solution was to bring the railway within the defences, in the same way that seventeenth- and eighteenth-century Dutch towns had incorporated the introduction of canals (and their associated harbours) within 'water gates'. It was felt by most military opinion, however, that the breaches needed in the fortifications would introduce a fatal weakness into the system, and that the railways should be kept outside, or discouraged altogether. However, extra-mural

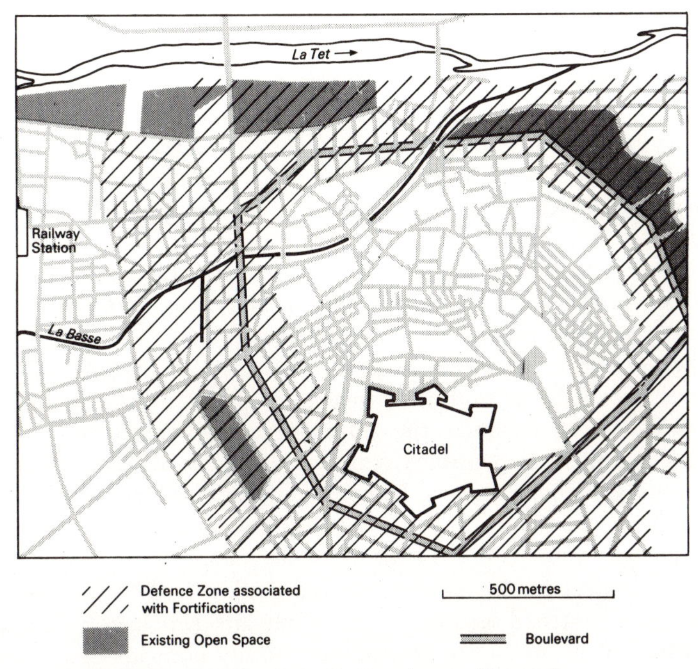

Figure 2.21 Fortifications and land use in Perpignan (France)

locations generally obstructed fields of fire, and in many European cities this dilemma was the catalyst that led to the ultimate dismantling of what had become semi-obsolescent defences – whose maintenance would have involved an unacceptable rejection of the new railways. Where the defence function was still important, other solutions had to be found. In Perpignan, for example, on the French–Spanish border (see Figure 2.21), the military authorities required that the railway remained at least 700 metres outside the fortifications, thus beyond the glacis of around 400 metres; and that the railway station was to be constructed of prefabricated sections that could be easily dismantled in time of war so that fields of fire could be maintained (Ashworth and Schuurmans 1982).

One of the few aspects of the effects of defence needs upon urban form that has received consistent attention in the literature is the influence of fortifications upon street and block patterns. What might be termed the conventional wisdom on this point was expressed as early as 1911:

> As a general rule the walled towns were much more regular in the planning of their streets than those towns which were always open; this is more especially the case in those which received their mural girdle early and particularly in towns of Roman foundation.
>
> (Harvey 1911)

Unfortunately, this set of simple generalizations, which has been regularly reiterated since, does little justice either to the variety of military demands or to the variety of individual urban reactions to them.

The simple rectangular grid geometry of the Roman walled town seems to have captured the imagination of historians by its ubiquitous reproduction in a recognizable standard form throughout the empire. Its regular patterning of streets based upon the two main intersecting highways (*cardo* and *decumanus*) frequently survived in the successor medieval towns, and remains recognizable in the contemporary street pattern in many towns in many countries (see, for example, Aston and Bond 1976 for the Roman origin of the street plan of a number of British cities).

However, it is by no means so clear that the well-known Roman standard design is principally a result of defence needs. A rectangular circumvallation does not, in fact, optimize the length of circumference to be defended compared to the area protected within it. It also accentuates the awkward problem of vulnerable corners. Even more significant is the fact that the grid street-pattern (based on straight main roads linking gates placed in the centre of the curtain walls) is perhaps the least defendable arrangement. In other words, defenders cannot manoeuvre and deploy easily within the city to defend any threatened section of wall, while attackers have a clear uninterrupted field of advance into the centre of the city once any gate is breached.

A better explanation of the Roman model probably lies in the administrative and organizational structure of the Roman army that invented it. The wall was intended as little more than a picket line – enclosing accommodation for a garrison which would take to the field in the face of attack and not defend a wall. Rectangular simplicity and grid street-patterning were thus a result of the unimaginative application of standard practice by the military engineers for housing a military organization in barracks (drawn up in blocks with a parade-ground precision) so that the 'town and the encampment became closely allied' (Carter 1983: 28). In fact, the Roman town has many of the characteristics of the military cantonments of imperial India, 2,000 years later – as a result, no doubt, of similar needs to accommodate garrisons and their ancillary services, which were physically laid out according to the customary functional and social military hierarchy (King 1976).

The textbook idealizations produced from the fifteenth to the seventeenth centuries provide better examples of street- and block-patterning being determined by military requirements. Most of the Italian renaissance models (described earlier) attempt to exploit the advantage of 'interior lines' and central deployment spaces in various often ingenious, ways – including broad radial avenues, circumferential ring roads, and even a spiral road in the case of Martini's 1451 plans for hilltop towns. A minority, contrasting and somewhat pessimistic view (Maggi's 1564 plan) even favoured a system of narrow alleys that hampered the progress of an enemy once the wall had been breached (see Argan 1969) – as opposed to broad boulevards along which reinforcements could be moved within the city. However, it is really only in such optimal models, and in a few fortress towns planned and constructed according to such blueprints (e.g. Palma Nova, Naarden or Neuf Brisach) – in which defence requirements took precedence over other urban functions – that such patterns can be recognized. Most fortification systems were constructed around existing towns, and few towns were constructed or reconstructed within existing fortifications. Ironically, the most important effect that fortifications had upon street-patterns occurred as a result not of their construction but of their dismantling, which will be described in detail in Chapter 7.

Walls had effects upon the functioning of cities as well as their form. The wall was the actual, as well as symbolic, limit of civic law and civic order. Its purpose was as much to allow the policing of the city through the combination of wall, curfew and watch, as to defend it from external attack. The provision of a means of controlling and registering those passing in and out of the city allowed fiscal and security control. The walls served as a customs barrier, occasionally being built principally for this purpose (a *mur d'octroi* rather than a defensive *enciente*). Municipal finance in many cities throughout Europe was dependent upon *murage* (the tax on goods passing through the walls), which was originally intended to provide for the walls' upkeep. It is not surprising that the city's walls became a symbol of urban government, law and urban freedoms. 'Historically, legal systems for enforcing property and personal rights emerged from within the walls of cities' (Bish and Nourse 1975: 190). Those residing outside these limits (whether by choice or compulsion) were 'outlaws' – outside the obligations and privileges of urban taxation, regulation and protection.

Outside the walls of medieval cities could generally be found those trades too dangerous (such as metalworking with its risk of fire) or noxious (such as tanning) to be permitted within. The post-medieval fortifications had large unoccupied areas outside them, and such building-free zones were attractive for activities needing either temporary occupation of space or access to the water of the water-defence systems. Many textile operations (such as dyeing) needed both, and were often even tolerated on the bastions themselves. This sort of functional zoning often corresponded to a social segregation where non-conforming social groups (whether ethnic minorities such as Jews, gypsies or other aliens, or just social misfits) were left to settle outside the walls. City gates thus became break-of-bulk points and, often, tax-free zones – as travellers and their goods were cleared into and out of the city. The district around the gates was an obviously advantageous location for transport-related trades (such as saddlers or blacksmiths), and for what today would be termed the 'hospitality industry', in its various branches from currency transactions to prostitution.

Such locational patterns were noticeable in the Roman distinction between *oppidum* within the walls and *vicus* beyond, and continued in the medieval town in the contrast between *bourg* and *faubourg* (Vance 1972: 120). The freedom from civic and guild restrictions, and the presence of enterprising social groups excluded from the main conventions of urban society, created what were in effect 'free enterprise zones' which frequently developed into the major trade, manufacturing and entertainment districts of the city, recognizable long after the gates had been dismantled.

Figure 2.12 shows an example typical of the Dutch fortified city, where a dependence on water transport accentuated the concentration of commercial activities in the 'harbours' (or inland, water-transhipment points) just within the 'water-gates'. Consequently, the main commercial activities of the city were dispersed (until well into the middle of the nineteenth century) through a ring of sub-centres composed of shopping and commercial-service activities grouped around these harbours.

Thus the fact that the city was walled, often continuously for a thousand years, could not fail to have an impact upon its growth, form and functional structure. Some of these effects were known and welcome, others were neither, but both are frequently still to be recognized in the patterning of the city.

Chapter 3

The defence town

DEFENDED TOWNS AND DEFENCE TOWNS

The previous chapter has considered the effects of the long history of urban fortification on the form and functioning of towns: this chapter will survey a number of towns in which the defence function is so significant that they can be collectively labelled 'defence towns'. The two categories overlap, in that some of the fortified towns already described were also 'fortresses' in that they formed part of defensive strategies far wider than the protection of the town itself, and it is clear from many of the examples that some towns possessed defensive systems far more elaborate, as well as costly, than could be justified by their defence alone. Equally, some of the 'defence towns' described in this chapter were themselves fortified, but many others were not. The distinction is that while defence, in one form or another, is a factor in almost all towns, there are some in which it assumes such an importance – either relative to other local urban functions or in relation to wider regional, national and international considerations – as to justify their consideration as a distinct category. Lotchin (1984) coined the term 'martial metropolis' for a town so 'moulded by its alliance with the military', in its origin, character or functioning, as to need separate treatment.

DEFENCE INDUSTRY TOWNS

The discovery of the existence of the military–industrial complex was soon followed by the definition of its urban component, the metropolitan–military complex. Both were initially traced as responses to the growth of defence-equipment-procurement spending in the United States – between 1941 and 1945 ('the arsenal of the democracies') and from 1950 (Korean War rearmament) through until the 1980s (the arms race with the Soviet Union).

Studies of the distribution of military spending at the national scale have been produced for, among others: the United States (Stein 1985); the United Kingdom (Short 1981); and West Germany (Kunzmann 1985). The role that such spending plays in regional economies – and (consciously or not) in regional development strategies – has also been argued in various contexts (Todd 1980; Markusen 1985). It is a short logical step to proceed to consider the effects of defence industries upon particular urban examples – such as the Norfolk naval base (Silver 1984); the shipyards in Vallejo, California (Schneider and Patton 1985); the aerospace industries of Seattle (Abbott 1984) or Bristol (Lovering 1985); and the small-arms-manufacturing towns of New England (Lotchin 1984). The very diversity of such towns makes generalization difficult. However, they do possess a number of common features as a result of the defence connection.

A major feature of such towns is that the defence function involves a dependence upon trends and decisions existing beyond the control or influence of the town itself. The town exists in the service of a defence system that operates on a wider spatial scale. Its economic prosperity and even continued existence depends upon strategies and fortunes determined elsewhere, and quite small adjustments in national policy or international relations will be magnified on the local scale – resulting in abrupt fluctuations in the urban economy. Consistency in demand has never been a feature of defence spending, which by its nature is episodic, reflecting the variations in international tension, political priorities of governments, and technological change. National defence policies will dominate local economic planning, and local political decision-making will not only be subordinated to such policies but will frequently be reduced to lobbying at national level to influence such policies in the economic interests of the city. Thus the economic dependence will be reflected in a political dependence.

The monofunctional nature of such an urban economy has frequently been encouraged in the past by the military authorities, in order to preserve the effectiveness of the garrison function by precluding any potential competition of other possible functions for space or manpower. A number of aspects of this persistent, deliberate exclusion of non-military functions in order to preserve a local monopoly of both skilled labour and use of harbour facilities for the British Royal Navy has been shown by Riley (1987) in the example of Portsmouth – which he describes as having possessed many of the economic characteristics of a 'company town'.

The consequences of this dependence upon defence industries is most starkly revealed when shifts in national or international defence requirements reduce or remove completely the defence function. Retreat from empire (whether Roman, Venetian or Portuguese) left a string of functionless defence towns. Somewhat more recently (and more rapidly), the British withdrawal from 'East of Suez' has presented problems of civilian conversion to numerous dockyard towns and naval bases. Some (such as Trincomalee or Simonstown) substituted a national for an international role; in others – such as Malta (see Elliott 1982), Gibraltar or Singapore (where 25 per cent of the total GNP as well as 10 per cent of the total national land area, was accounted for by the naval dockyard when it was closed in 1971) –

the defence function was the *raison d'être* of not only the town but also the country in which it was set, and radical reorientations were needed.

An evaluation of success in conversion of urban economies from a defence dependence to other activities would go beyond the scope of this book, but the attempts to use the legacy of former defence functions (whether in the form of space, manpower skills, or symbolic associations and historical relics) will be considered later in Chapters 7 and 8.

This economic dependency is easy to recognize with hindsight, and to deplore its consequences in the case of the ship-building and maintenance facilities of the naval base, but it is equally evident (although rarely so condemned as a hostage to future economic fortunes) in the case of modern air-force bases and their relation to high-technology defence industries. Air bases have proved attractive magnets to defence industries as a result of the need for face-to-face collaboration between contractors and customers. In addition, there is the possibility of local recruitment of ex-service personnel into defence firms (Markusen and Bloch 1985). The result has been the development of industrial complexes around major air bases, particularly in the United States: for example, around Edwards Air Base in Southern California; in Los Angeles and Orange counties (where almost half of all manufacturing employment was in defence-related industries by the mid-1970s); in San Antonio (which has more military bases than any city in the US); or around the Cape Canaveral/Patrick Airforce Base complex in Florida.

However, not all the characteristics described above necessarily apply to all towns with military functions, and some military facilities have either little impact on the urban scene or have more in common with other non-defence functions. Many defence-equipment industries differ little in their locational choices (and urban impacts) from similar industries serving civilian markets. Furthermore, urban economic dependence and vulnerability to shifts in demand are not confined to defence-related industries. For example, most aircraft or ship construction has similar characteristics, whether the orders are military or civilian. Military academies have much the same impact upon towns – such as Breda (The Netherlands), Kingston (Canada), Sandhurst (UK), or Annapolis (US) – as any other large institute of higher education. Similarly, as Wells (1987) has traced in detail for the United Kingdom, the locational concentration of defence-related research establishments is only in part related to the garrison towns and military bases; they are as much explainable in terms of access to centres of government, higher education and research, and advanced industrial complexes – in much the same way as similar civilian research establishments. Even the administrative offices of armed forces, such as the Royal Naval headquarters at Bath, were selected for similar reasons (and play much the same role in the town) as the administrative offices of any large organization.

THE GARRISON TOWN

It is obvious that the fortified towns discussed earlier also accommodated garrisons to man their defences. However, an important distinction can be drawn between a town with a garrison and a garrison town. In the first case, a military force (whether composed of professionals or local militia) is present on a permanent basis (or raised in time of imminent danger) in order to defend the city, while in the second case the city exists in whole or in part to provide services to the military in fulfilling a wider strategic garrison function. Thus, although all garrisons by definition are engaged in defending places (including the towns in which they are stationed), the idea of the garrison town is rather more specialized, implying that the stationing of such garrisons has regional or national purposes. Such a garrison town – and the term is used broadly to cover the stationing of all military forces whether for land, sea or air defence – may or may not be fortified, but the defence of the town itself is not the primary purpose of the presence of military forces.

The characteristics of garrison towns

Armed forces, if they are to be maintained on any sort of permanent footing, need space. Many of these land-use requirements are not only most likely to be found in existing cities but may in themselves be a factor initiating urban development. Space is needed: for training and mustering; for the storage, maintenance and repair of arms and equipment; for the dumping of caches of expendable supplies of food and munitions; and, above all, for living accommodation for personnel, their families and those providing services to them. Even the simplest equipped and organized militia requires an open space or drill hall to gather and parade, as well as a secure weapon store or arsenal. These land-use demands have increased in amount and have become more specialized as a result of many of the historical changes discussed in Chapter 2. Not only have forces grown larger but they are maintained in peacetime – therefore requiring personnel to have separate, segregated and permanent accommodation rather than living at home or being temporarily billeted on citizens. Forces have become equipped with weaponry too sophisticated and expensive to be the private property of the soldier, and too specialized to be mere adaptations of civilian artefacts, thus requiring specialized production and maintenance facilities; new forms of transport have been acquired which themselves need maintenance as well as supplies of food or fuel; and tactics and equipment have been adopted that need longer training to master and a permanent organization to administer.

All of these long-term trends, operating over a period of centuries, have increased the importance of defence forces as a land use and as an economic function in its own right. These requirements for barracks, supply depots, arsenals and training areas formed the justification for a specialized form of settlement, most easily described as the garrison town. Some of these functions (such as training in the manoeuvring of large units or the use of dangerous equipment) are best accommodated in rural areas, while others (such as supply depots) can be located outside towns. Most functions are inherently urban (in which case they require access to the support facilities of cities, including their centrality within transport networks). However, the military requirements may be so extensive that they become the *raison d'être* for the existence of new towns, whose functioning and morphology is shaped by this activity. In either event, the garrison town is a specialized type of urban settlement whose long history and distinctive characteristics deserve more attention than has been given in studies of the morphological evolution, economic functioning and social and political geography of towns.

The nature of the functions that such towns fulfil endows them with a number of distinctive characteristics. First, and most fundamental, can be the very location of the town itself – which may be chosen for its place in a wider strategic setting, or for its possession of local site characteristics valuable to the defence function. These locational attributes may fortuitously also be those required for other functions, enabling a town such as Singapore or Cape Town (originally located for defence purposes) to diversify and grow on the basis of other activities. It is, however, equally possible that the particular locational requirements of defence are quite different from those of most other urban activities. In this case, the disappearance of the defence function threatens the continued existence of the town. Very few Alexandrian, and only a minority of Roman, military settlements survived the empires whose frontiers they were designed to defend. Naval bases, such as Gibraltar, Halifax (Nova Scotia), or Den Helder (The Netherlands) – whose siting was ideal for prevailing imperial strategies and marine technology – find themselves in decidedly sub-optimal locations for the development of commercial relations with their immediate hinterlands. Similarly, it may be necessary for security reasons to place garrisons in areas where other urban functions are unlikely to develop autonomously: for example, the military bases of the British in the Scottish highlands in the eighteenth century; the French in the Sahara in the nineteenth century; or the Americans in the Arctic in the twentieth century.

Second, the garrison function also involves a similar dependence upon external decisions (as discussed earlier for defence-industry towns): a situation often underlined in this case by the subordination of local representative authorities to military governors. The monofunctional nature of such an urban economy (with its concomitant, inevitable instability) may be accentuated by the peculiar locational attributes of garrison towns (as mentioned above) which make diversification difficult.

Third, this economic and political dependency can also be reflected in the local urban society – which can become dominated both by military personnel themselves and also by the attitudes and values held by the military. Where these differ significantly from national norms, then the garrison town takes on many of the characteristics of a colonial implant – separated economically and socially from the immediate hinterland in which it is set. It is not only the serving military personnel themselves who may have a decisive influence upon the social structure of garrison towns. Civilians employed in defence-related activities (or retired from military service) may contribute to the shaping of quite distinctive social patterns. Harris (1987) has traced in detail the characteristic social ecology of British naval-dockyard towns in the nineteenth century, using especially the example of Sheerness. Similarly, but using more recent data, Jones (1987) has demonstrated the numerical significance of military married quarters – through their concentration at particular locations in England and Wales – and thus the importance of their influence upon the social attributes and attitudes of particular towns.

Fourth, a characteristic that stamps a clear visible tone on garrison towns is the sheer physical presence of military personnel, and the consequent provision (officially and unofficially) of facilities to serve this large and distinctive consumer market. Officially-provided barracks and canteens are a relatively recent development, resulting from a combination of the difficulties of local billeting together with increasing concern for the health, welfare, and moral well-being of troops from both official and charitable organizations. These were later supplemented by estates of married quarters (for example, the refortification of seventeenth-century Copenhagen by Christian IV included one of the earliest purpose-built complex of married quarters, at Nyboder). These, and the related social and sporting facilities that were successively added, can form a substantial and dominating element in the urban architectural scene. The private sector, by responding to the opportunities of a captive market of large numbers of young men away from home, and with generally long periods of leisure time, has traditionally provided entertainment. Garrison towns typically have a higher than average number of bars, night-life and prostitution establishments, with some districts – such as Valetta's Strait Street ('The Gut'), Singapore's Malay Street, Middle Road and Dhoby Ghaut – achieving an empire-wide notoriety in these respects.

Fifth, there is a less visually dramatic but possibly longer-lasting influence of the presence of a garrison on the functioning of the town – namely, its role in the genesis of a tourist industry. The recreation needs of the military could play a role in preserving recreational open space and stimulating recreational demands that, occasionally, can provide an inheritance of facilities of use later for tourism. The enlivening effect of the presence of garrison officers upon local society, and even the retirement of pensioned or half-pay officers to familiar high-amenity districts near the military establishments, could provide the stimulus for tourist-resort development. Riley

(1972) has traced these relationships between the naval base of Portsmouth and the adjoining seaside resort of Southsea, and Hedges (1973) has shown how even a quite small garrison stationed in Yarmouth during the nineteenth century was instrumental in the stimulation of recreation development on the sea coast outside the town, which later became the core of the tourist resort. In Malta the development of the resort function of Sliema has a similar origin (Bradford 1985).

Last, there are the characteristic morphological features of garrison towns. Here it is far more difficult to generalize about common features of urban form. The necessity for mass communal sleeping, eating, exercising and equipping has architectural consequences in the large regimented buildings and spaces that are a typical feature of such towns. The obvious basic functional requirements, together with the hierarchical structures of armies, produce very similar patterns – whether in Roman legionary bases, or British–Indian cantonments (King 1976). Similarly, the separation of provision for the garrison from that for citizens, by barriers or spaces, was encouraged by the fears of either the citizens of the misbehaving military, or the military authorities of contamination by civilian ideas and vices. However, garrison towns may or may not possess fortifications. In practice, many did not, the presence of the army being deterrent enough, and many were only minimally fortified, as much to keep the empressed army in as the enemy out. There is a modern trend for increased 'fortifications' in the form of various physical and electronic barriers but these are intended not as a defence against enemy forces but against local espionage, terrorism and political protest. At the other extreme there are occasions when military installations have been strongly fortified, and the clearest (and frequently most spectacular example) of this situation is the special case of naval bases, considered later.

The contemporary garrison town

Britain, and its defence needs overseas, may provide the best historical examples of the naval base but conversely presents few good examples of the effects of land-based garrisons on the urban structure, as a result of the absence of a tradition of a large standing army and the stationing and training overseas of a large proportion of the small professional army for the last 200 years. Germany, on the other hand, has an equally long tradition of maintaining within its borders large numbers of land forces, whether German, allied or occupying enemies. The amount of land required for the direct needs of such forces was calculated by Sicken (1977) to have totalled 386,000 hectares in 1939. By 1972, contrary to what might be expected in a peacetime Federal Republic little more than half the size of the immediate pre-war Reich, the requirement had risen to 423,000 hectares, as a result of the continuing strategic importance of the German land corridor (reflected in its title of the 'Central Front' in the global east–west confrontation) and the stationing in it of both large, locally-raised forces and allied contingents.

Hofmann (1977) has recognized three types of garrison town in the contemporary Federal Republic, each of which has its distinctive requirements and planning considerations.

1 What Hofmann terms 'ordinary' garrisons of either the regular army (*Bundeswehr*) or locally-raised territorials (*Landwehr*). Both require access to road and rail networks for procurement and deployment, the former in centres with good national accessibility (such as Stuttgart, Krefeld or Erlangen) while the latter is to be found in regional centres.
2 'Allied garrisons' have similar requirements for local urban services with a special need for international communication with the home country. The long continuous experience of accommodating a foreign garrison from the same country leads to the emergence of strong economic, cultural and matrimonial links between the garrison and the locality, despite inevitable frictions. In West Germany towns such as Frankfurt or Nuremberg have become in some respects distinctively American (US Army Fifth and Seventh Corps respectively) while Rheindalen or Hamelin (BAOR) are similarly British. Berlin has hosted four foreign garrisons since 1945, each of which has imposed its character on a different sector of the city.
3 'Frontier garrisons', whose task is partly border policing and partly 'trip-wire' invasion alarm, are necessarily widely dispersed in small units stationed mostly in small towns, especially along the former East German and Czechoslovak frontiers (such as Bad Kissingen). Their location creates special access and transport difficulties, especially in areas such as the Harz mountains.

All three categories may be superimposed upon the same region, with units of different size and function being accommodated at different levels of the urban hierarchy. For example, Figure 3.1 shows the territorial disposition of the US Fifth Corps in 1980, which was charged with the defence of the Fulda Gap as part of central army group (CENTAG) on the central front in West Germany. Corps headquarters, with its concomitant administrative and support personnel, are located in the major urban centre of Frankfurt with its international transport and communications functions. Brigade and regimental headquarters are located lower in the urban hierarchy at Friedburg, Gelnhausen and Kirchgoens, with forces at battalion and even company strength being dispersed in the smaller settlements (such as Bad Hersfeld and Fulda), some of which are operating as frontier patrols. In addition, German territorial units (*Heimatsschutzenkommando* and *Wehrbereichskommando*) organize their unit headquarters and mustering locations on the basis of the settlement hierarchy.

When training was little more than the drilling of small units, it was

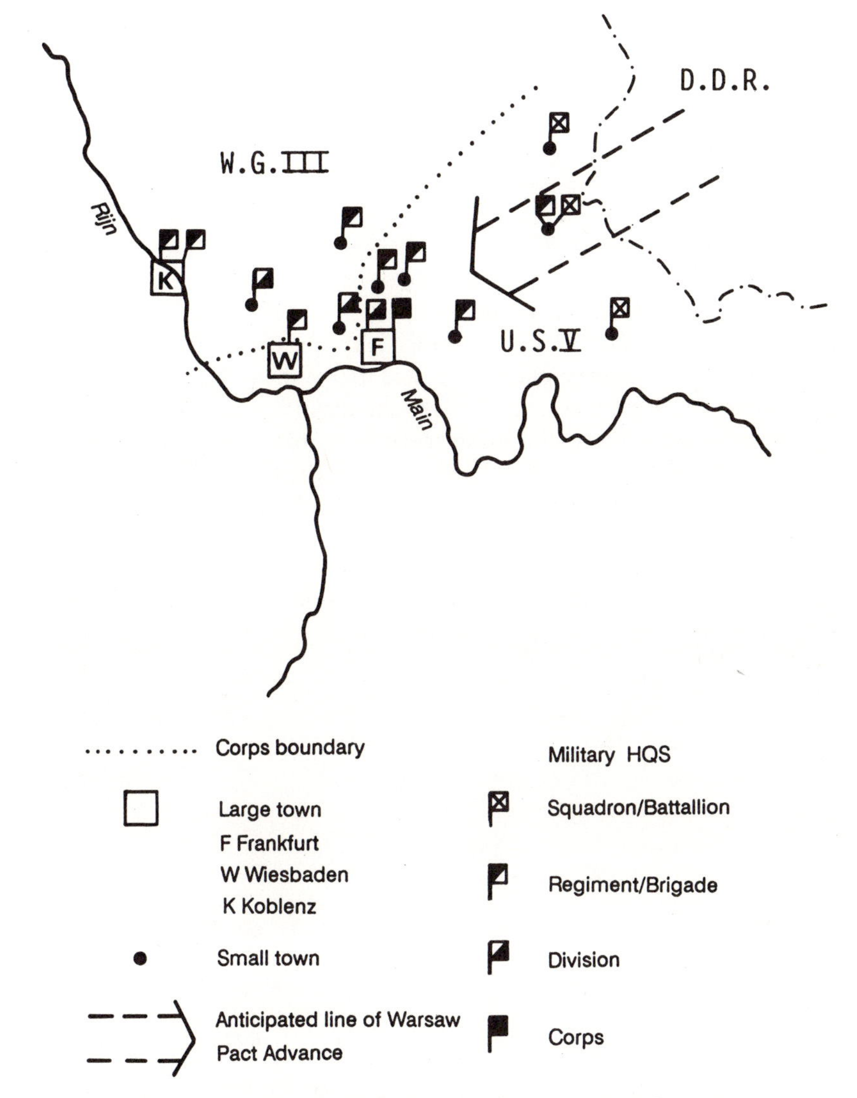

Figure 3.1 The garrisoning of US Sixth Army Corps in West Germany

accommodated within the city, and the parade ground was a conventional urban land use with the spectacle of military parades a common urban entertainment as well as a demonstration of prevailing political sentiment. The requirement for more space for the manoeuvring of larger and mechanized units and the firing of larger calibre weapons led to the migration of training areas out of the city to less densely inhabited areas. Munition, fuel, heavy-weapons and vehicle depots have tended to follow this outmigration. This has occurred in part as a response to pressures from local communities and their planners who view such land uses in densely inhabited areas as undesirable, and – when potentially dangerous munitions (up to and including theatre nuclear weapons) are being handled – clearly unsafe. In part also it was convenient to locate such equipment near the urban fringe areas where it was to be exercised. This has in turn led either to the separation of the inner urban barracks from their depots and training areas or, when the opportunity has arisen, for the construction of new barrack complexes on the urban fringe. In a number of West German cities the curious situation has arisen that the allied occupying forces took over the existing Wehrmacht barracks in 1945, most of which were constructed before the First World War in inner city locations. Such units now find themselves inconveniently separated from their depots and training sites, while the West German Bundeswehr mostly occupies more recently built accommodation on the urban periphery, often with good access to the post-war transport network. Figure 3.2 shows an idealized pattern of the location of military installations in a medium-sized West German town.

These locational patterns give rise to two serious defence problems. First, the sort of military training experienced is constrained by the problem of

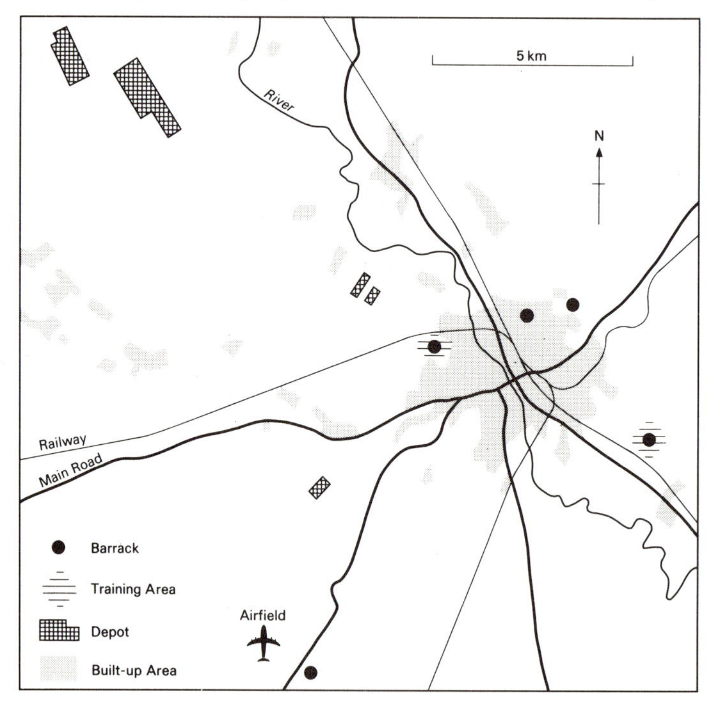

Source: Sicken (1977)

Figure 3.2 A model of the location of military installations

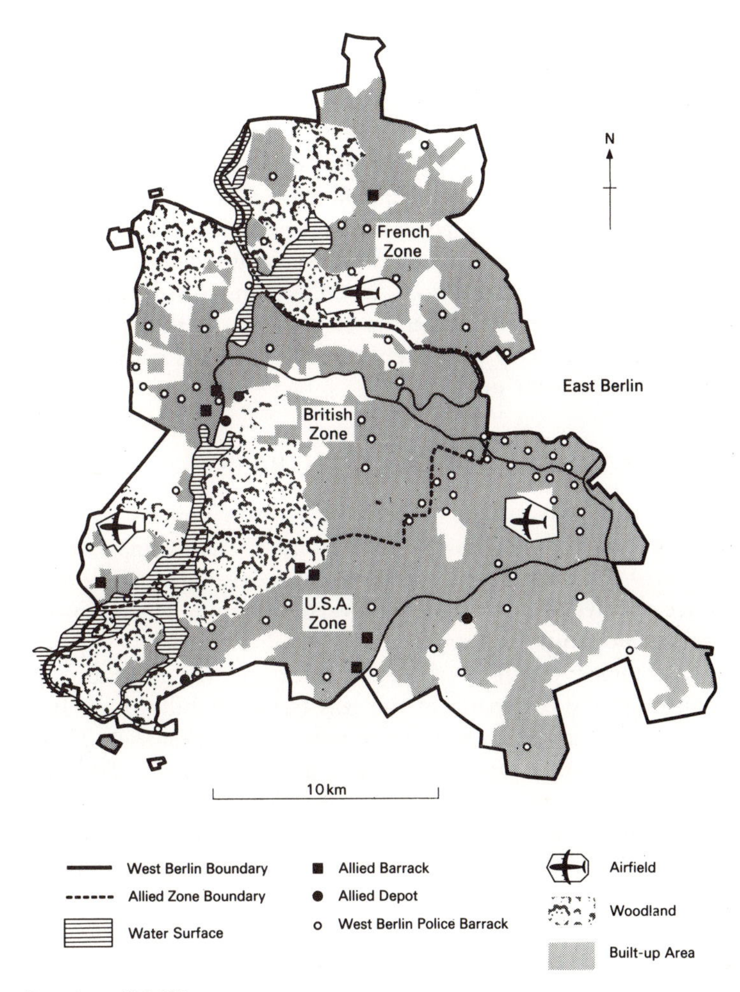

Source: Dunnigan (1980)

Figure 3.3 Military installations in West Berlin

accommodating such training, in multifunctional towns and urban regions, within democratic societies in peacetime. In particular, most training now occurs almost exclusively in rural areas whereas military services are most likely to be required in urban environments, either in support of the civil power (Chapter 4) or as the most likely battlefields in future global conflict (Chapter 5). For most contemporary and anticipated military tasks, no weapons or tactical training that approximates to the reality of urban warfare is possible. Second, the peripheral and scattered location of military installations around the city, and the spatial separation of barracks from depots, handicaps the capacity of defence forces to react to emergencies. West Berlin, for example (Figure 3.3), is a city that was physically surrounded by a potential enemy from 1945 until 1990, and was particularly vulnerable to overrun with conventional forces. It was therefore heavily garrisoned by West German military police and allied units from three countries. In particular, the dispersed and peripheral locations of the British and US garrisons would have made co-ordinated defence and rapid reaction difficult (Dunnigan 1980). A more general point was made by Kamps (1980), reporting that US units on the Central Front required 48 hours from alert to ready status because of the need to collect and distribute larger-calibre ammunition from peripheral depots as a result of the spatial separation mentioned above, a delay also likely to have occurred with British but not necessarily Bundeswehr units.

THE SPECIAL CASE OF THE NAVAL BASE

Although not all naval bases necessarily contain military garrisons, their purpose by definition derives from beyond the cities in which they are located. They are, therefore, examples of cities existing in whole or in part in the service of defence rather than the reverse situation, which is more usually the case (see previous chapter).

The far-ranging mobility of naval forces and the flexibility that they confer through the command of the seas has an Achilles heel. Fleets need fixed permanent land bases for their supply and maintenance. These in turn are vulnerable to blockade by sea and assault by land. The greater the dependence of a state's policies upon thalassocracy, the greater the care and finance lavished upon the protection of the bases upon which this power rests. Historically, the maxim was simple: trade and prosperity depended upon the fleet but the effectiveness and security of the fleet depended upon the location and security of its bases (Mahan 1890). Examples of this can be found from the building of the 'Long Walls' around 470 BC to link Athens to its naval base at Piraeus, through the flowering of the various sea-route-oriented European trading empires (e.g. those of Venice in the fourteenth, Portugal in the sixteenth and Holland in the seventeenth centuries), all of which were in fact little more than strings of fortified naval bases from which the fleets could exercise control over the sea routes between.

It is therefore not surprising that the containment by blockade of an enemy's naval bases (such as the Royal Navy's off Brest and Toulon 1793–1802, 1803–15; or the US Navy's off the Confederacy 1861–5) has been a common naval strategy. Even more desirable was their temporary seizure by surprise raids (as on a number of occurrences, two of which were successful, on Portsmouth by French ships from the fourteenth to the sixteenth centuries, on Chatham by the Dutch in 1667, or on Zeebrugge by the Royal Navy in 1918); or, if possible, their permanent occupation by seaward assault (as in the case of New Orleans and Mobile 1864); or long-drawn-out land siege (as with Port Arthur 1905–6, or Sebastopol 1854–5). In practice, attacks on the enemy's naval bases have been the more usual technique for attaining naval success than the set-piece battle of opposing fleets at sea.

History's most complete example of a world-wide, seapower-based empire – eighteenth- and nineteenth-century Britain – provides some of the best examples of such defended bases, in a global network (such as Halifax, Antigua, Bermuda, Singapore, Gibraltar, Malta). In home waters, the location of British fleet bases was largely a response to the prevailing enemy: for example, the east coast against Holland (the Medway and Thames bases) or Germany (Scapa Flow, Rosyth, Harwich); the south coast against France or Spain (Portsmouth, Portland, Devonport). The defence of such bases provides an exception to the early abandonment of urban fortifications in Britain.

The common characteristics of such bases can be examined in more detail through some widely scattered examples – Portsmouth, Malta, Halifax (Nova Scotia), and Singapore – all of which were part of the Royal Navy's world-wide *Pax Britannica*. The first three are all located on fine natural harbours in which the naval base occupied only a small part, although expanding over time in response to changing maritime technology. The most complete, long-standing and best-described history is probably that of Portsmouth (Saunders 1967; Temple-Patterson 1967; Corney 1983), designated as a royal dockyard on the south-west corner of Portsea island in 1496, although an even earlier Roman base of the *Classis Britannicus* had been located further up the harbour at Portchester from the second century AD. A number of the many arms of Malta's extensive harbour have at different times been used as fleet bases but, in the most important period of maritime power, the Senglea isthmus between Frenchman's and Dockyard Creeks on Grand Harbour was used – and ultimately expanded after 1858 to include most of the isthmus (Hughes 1969). Similarly, the Halifax naval base occupies only a small part of the south shore of Halifax Harbour and the extensive Bedford Basin through The Narrows (see Figure 3.4). All four naval bases existed spatially apart from, although economically related to, residential and commercial areas. Portsmouth, Malta and Halifax were fortified, with successive lines of defence works being constructed and reconstructed in response to changes in artillery. Both Portsmouth and Malta were defended by walls and outworks covering the sea approaches in the sixteenth century, and again in the seventeenth century. In Portsmouth, curiously enough, it was the administrative and political core of the city that was fortified, rather than the dockyard itself (in Portsea), until the 1770s. In Malta, the area defended by fortifications was successively extended to

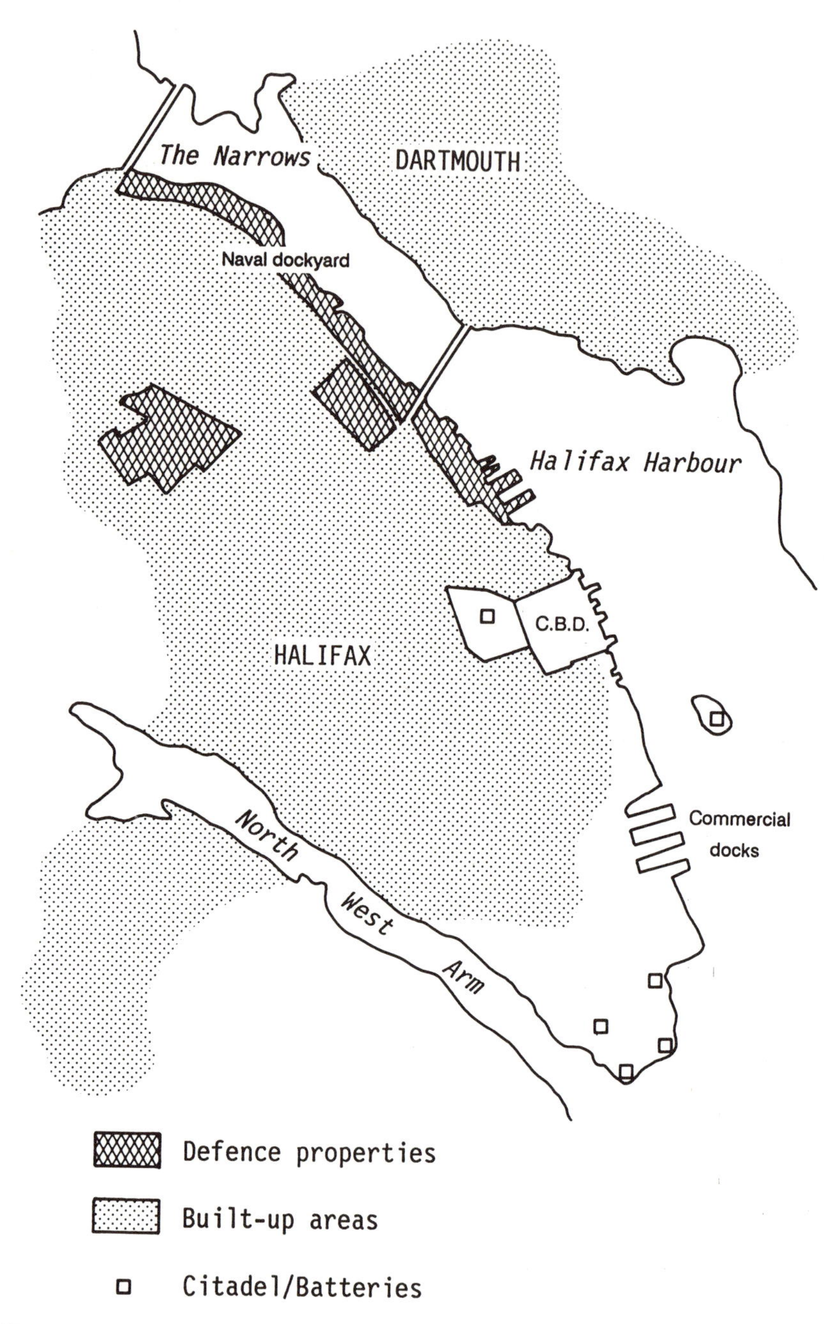

Figure 3.4 Military installations in Halifax (Nova Scotia)

include not only Valletta ('Floriana Lines') but also the 'three cities' south of Grand Harbour, with the 'Margherita Lines' (1638) and 'Cottonera Lines' (1670–80). All three bases responded to the revolution in the efficiency of artillery in the middle of the nineteenth century by the building of a completely new defence system based on artillery batteries covering the seaward approaches, and in the Portsmouth and Malta cases also the landward approaches (the 'Hilsea Lines' and later Portsdown 'Palmerstonian' forts, and the 'Victoria Lines' respectively).

The Singapore case, however, is significantly different. Although the island was occupied since 1819 as a result of the potential significance of its position on maritime routes, it had no imperial defence responsibilities before the opening of the Suez Canal. Some artillery batteries for local defence were sited to cover the commercial port. The naval base was quite specifically a product of British inter-war naval policy, and was begun in 1923, completed in 1941, and demolished almost immediately in the face of the advancing Japanese. Unlike the three other cases above, it did not evolve incrementally (i.e. with modern facilities being added to long existing ones). The deliberate decision was made to develop a greenfield site at Seletar on the Johore Strait to the north of the island, rather than on the crowded Keppel Harbour in the south, thus completely separating the naval base not only from the civilian town but also from the commercial harbour. Artillery defences were built on the north shore of the island covering the sea approaches to the base along the Johore Strait, and (contrary to popular mythology) faced north towards Malaya (Turnbull 1977).

A HISTORICAL DEVELOPMENT MODEL

It was on the basis of the Portsmouth experience that Riley (1987) constructed his model of the evolution of a fortified dockyard town (see Figure 3.5). Although intended as an explanation of land-use changes in naval dockyards, most of its characteristic structural elements – and the external factors effecting change in them – would be equally relevant to any fortified garrison town. The evolution through the three stages is powered by the changing land requirements of the dockyard – the central feature and *raison d'être* for the town – in response to technical advances in naval architecture, propulsion and gunnery. In turn, the needs of the dockyard for stores, services, labour and protection leads to the location of a range of facilities around the dockyard, the whole being contained within extensive fortifications. The extent of the dominance of available space by the defence function is dramatically demonstrated in Riley's calculation of land-use categories in the twin fortified towns of Portsmouth/Portsea in 1860, which corresponds approximately to the point of change between stages 2 and 3 of the model. At that time the fortifications themselves accounted for 43 per

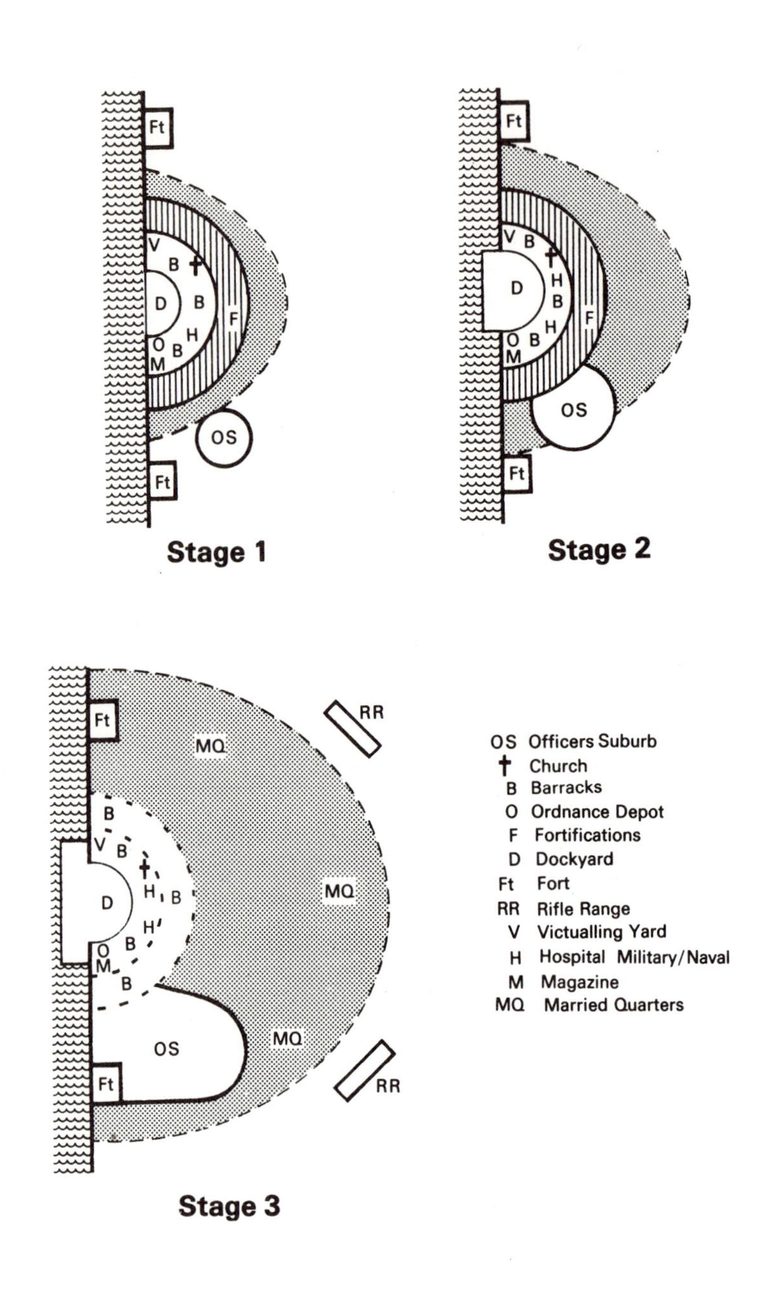

Source: Riley (1987)

Figure 3.5 Model of the evolution of a fortified garrison and dockyard town

cent, the dockyard a further 20 per cent and barracks and depots 11 per cent which left only 26 per cent over for civilian uses. The dismantling of obsolete fortifications and the release for development of both the land they occupied and their fields of fire allowed many previously constrained demands for expansion to occur. These included both direct military uses, such as weapons training, and also residential areas of military personnel, especially the 'suburbanization' of housing for serving and retired officers.

A garrison town, however, is also a town, and the only significant elements missing from the model are the ordinary ubiquitous urban functions, as 'civilian land-uses had perforce to mesh into a predetermined spatial pattern' (Riley 1987: 78). The commercial and administrative functions of the town will concentrate in the early stages outside but close to the dockyard near the centres of both economic demand and military decision-making. In the final stage, however, these civilian, commercial and administrative functions will tend to migrate inland in response to changes in accessibility, land requirements, and the shift in the economic centre of gravity inevitable with the expansion of all coastal towns. A new city centre is thus likely to arise on or beyond the former fortifications.

The adaption of such a model to garrison towns in general involves little more than minor modifications to the facilities and some changes in terminology. In an army base the victualling yard becomes the commissary, the ordnance depot the arsenal, and the barracks has added stabling. The only important differences are the absence of the dockyard (whose central dominance is not usually compensated for by parade grounds) and the absence in an inland town of the same pressure for the migration of the commercial centre.

Such modelling is a useful explanatory instrument, not only for describing historical change but also for suggesting some of the fundamental difficulties of adaptation inherent in such towns. The economic monofunctionalism is reflected in a distinctive morphological structure that remains as a major patterning element even after the garrison functions have been withdrawn or substantially reduced. The consequences of this evolutionary experience will underlie many of the re-use considerations described in Chapter 7.

PLANNED NETWORKS OF GARRISON TOWNS

Brief mention should be made of the planned establishment of networks of garrison towns as deliberate instruments of military and consequent political control of space because this has been (in a number of historical periods and over quite extensive areas) not only an important means of territorial extension of states and pacification of newly acquired areas, but has, in addition, frequently formed the basis for the subsequent development of enduring new settlement systems. Although such settlements have at different times been called by a variety of names, and have been established on different locational criteria and with different morphological characteristics, they share the common features of being deliberately sited by some central authority as interrelated systems rather than individual settlements – with the intention of fulfilling military functions.

Among the most well researched of such 'defence new towns' are the Roman *coloniae*, or settlements, of discharged army veterans, which were used to provide some defence in depth and trained reserve man-power in support of vulnerable stretches of the imperial frontiers (especially behind the middle Rhine, the Rhine–Danube gap of the *Decumates Limes* and on the Lower Danube). The influence of these upon the evolution of settlement patterns in these areas of Europe has long been the subject of study (Houston 1953; Dickinson 1961) but the Roman *coloniae* were only a continuation of previous Hellenistic policies in Mesopotamia and the Levant, where well over a hundred such settlements – the majority of which have not survived – were established for the same purposes in the hundred years after Alexander.

The use of such settlements to establish (first) military and (subsequently) political control was continued in medieval and early modern Europe. There are many well-known examples, including: the plantations of the *castella* in the *reconquistada* of Spain; the various systems of castle towns in the Norman conquest of the British Isles (see Graham 1988) and the securing of the Welsh borderlands; the *bastide* in the struggle for control of South-Western France between the Angevins and Capetians; the results of almost ten centuries of the sporadic German expansion through planned settlements eastwards in the *Drang nach Osten* across Prussia, Pomerania, Poland and Livonia, and the planned repopulation of the Hungarian Alfold and Danube Basin as it was reconquered from the Turks. However, one of the most carefully calculated and, at least theoretically, complete systems of defensive settlements was that of the ninth-century Saxon burghs, planned by Alfred and his successors: it was no less than 'a strategic plan designed to ensure the defence of England south of the Thames against military assault' (Blair 1962: 294). Thirty settlements – some of which were substantial, refortified towns of Roman origin (such as Winchester or Chichester) and others little more than villages – were walled and garrisoned through the expedient of allotting land holdings in a precise relationship to manpower needs. (The formula was: one hide of land supports one defender who mans one quarter of a pole of defence work. Thus Winchester, for example, with 3,300 yards of wall, was provided with a garrison of 2,400 men.)

However, the most notable system – in terms of the extent of the area involved – is that of the Russian fortified garrison towns, which were the effective instrument for the conquest of Siberia and Russian Central Asia, over some 300 years. Most usually a simple, easily replicable design of earth bank, wooden palisade with corner artillery bastions was extended (with few modifications) from the Oka River in the sixteenth century to the Pacific

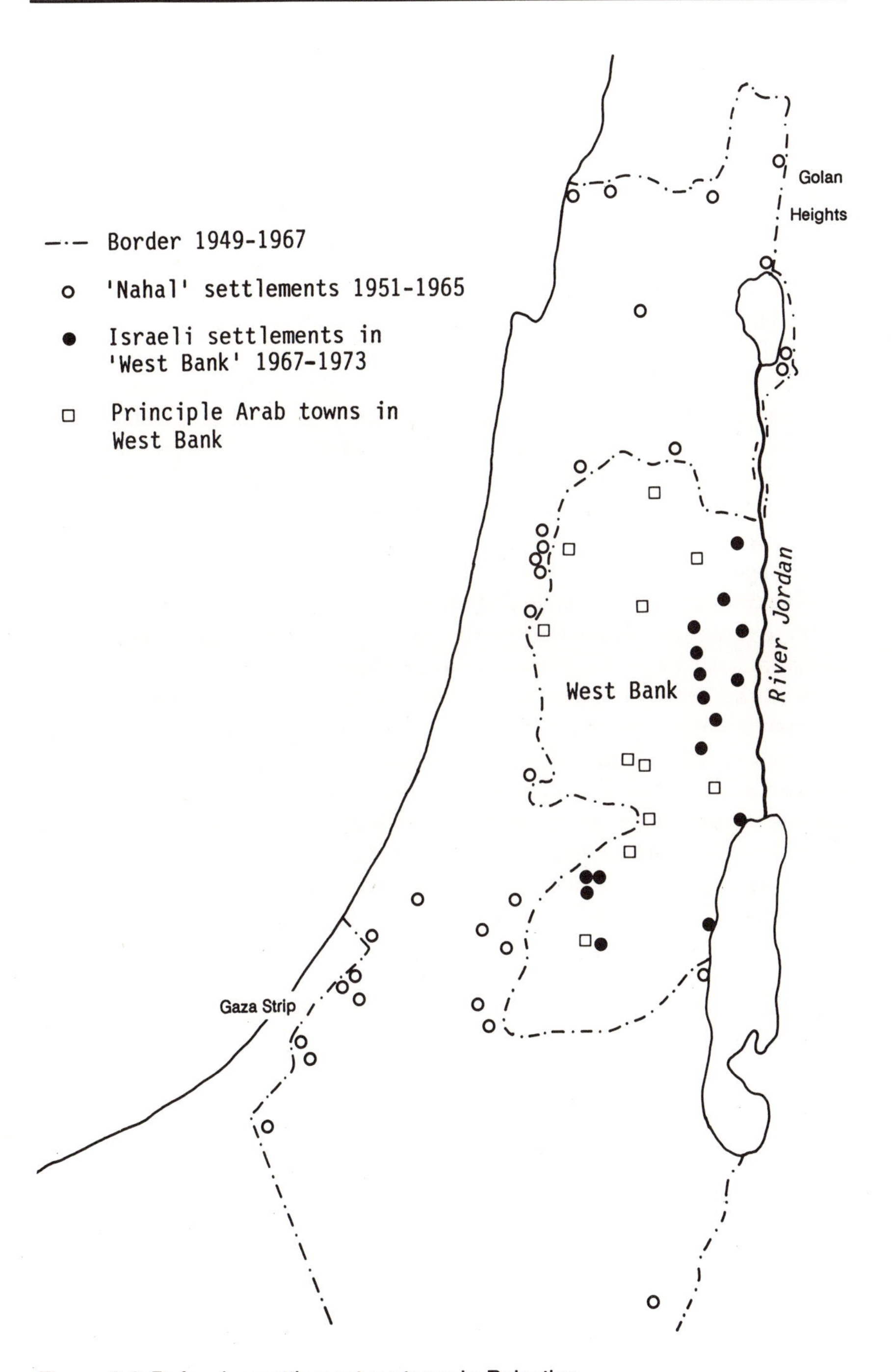

Figure 3.6 Defensive settlement systems in Palestine

Ocean in the seventeenth. Although military considerations were the primary locational factors and motive for such settlements, many rapidly acquired governmental, religious and market functions, some developing into major cities – such as Oriel (founded 1564), Voronezh (1630) and Orenburg (1743) (see Sumner 1944).

Such examples are not confined to the 'old world'. The struggle for dominance in North America between Britain and the United States resulted in a series of military settlements in Eastern Ontario between the St Lawrence and Ottawa rivers. British fears of further invasion from the United States after the war of 1812–14 led not only to the completion by the Royal Engineers of the Rideau canal in 1832 (bypassing the section of the St Lawrence vulnerable to US interdiction) but also to what amounted to a fortified defence line aligned along the canal from the main military base at Kingston to Bytown (later Ottawa) on the Ottawa river. This line included a series of regularly spaced, stone and wood block houses (among the best preserved of which is that at Merrickville) and the fortification of the canal locks and their associated buildings. This line was supported by deliberately established new settlements of discharged veterans who could provide a reliable trained militia reserve, and ultimately a loyal population. The largest of these traditional *coloniae* is Perth, laid out by military surveyors in 1816 and populated by veterans from Scottish highland regiments.

In more recent years, planned networks of defended settlements as part of military strategies can be found in the pacification programmes of what has become the conventional wisdom of anti-insurgency operations. Connected networks of protected settlements are inserted as islands of government control into insurgent areas – either defensively to separate existing populations from insurgents or aggressively as a means of extending control over areas – as used by the British in South Africa (1899–1902) and Malaya (1950–3) and by the Americans in Cuba (1898) and Vietnam (1965–75). These were generally small settlements and intended as much for local security as offensive operations.

The planned settlement policy of the State of Israel, however, has been both more comprehensive and has longer-term objectives. The communal agricultural settlements known as *kibutzim* proved to have military value as defence points against lightly armed opponents in the war of 1947–8 and the idea was extended in the *Nahal* settlements, of which thirty were established between 1951–67 under defence forces' control along vulnerable frontiers (see Figure 3.6). The occupation after 1967 of extensive new territories on the west bank of the River Jordan led to sixteen new settlements between 1967 and 1973 (Gilbert 1976). These settlements provide a source of armed manpower, a defence in depth of a vulnerable frontier area and islands of cultural and political control in the midst of a potentially hostile population, thus continuing a tradition of the use of such settlements as part of similar policies in that area which is over 2,000 years old.

Chapter 4

The insurgent city

It is difficult to draw a clear distinction between defence against external threats and that against internal insurgency. Such distinctions have rarely been drawn in history. City walls were intended as much to keep citizens in, and accounted for, as to keep enemies out. The wall (as has already been noted in Chapter 2) had a practical, as well as symbolic, jurisdictional purpose, enabling the urban authorities to exercise a control over the movement of goods and people, and thus served police, customs, fiscal and immigration purposes (as such flows could be channelled through the limited number of gates which would be opened and guarded at specific times). Thus the distinction between police and military structures was generally blurred, with the same forces being called upon to perform both functions. Indeed, the only distinction between the modern situation – especially in countries with the Anglo-American aversion to paramilitary police forces (Gellner 1974) – and that which prevailed in most countries until the last century, is that the military now operate 'in support of the civil power', whereas previously they were frequently the only effective instrument of that power.

External aggression and internal lawlessness have frequently been related – either as causes and effect, or as combined strategies in an assault upon the city. For much of the Middle Ages, the technical superiority of urban defence over urban attack determined that a strongly walled city was more likely to succumb to internal subversion, or a breakdown of internal order and resolve among the defenders, than to a direct assault on the fortifications from outside. General Franco's boast of a fifth column inside Madrid to supplement his four advancing upon it in 1936, was an echo of the hopes of most besieging forces during the preceding thousand years.

Although internal and external defence have frequently proved to be indivisible, it is nevertheless useful to consider separately the internal security of the city and attempts to subvert it through insurgency. The propositions underlying this chapter are that cities have proved especially important to insurgency, providing a particularly 'favourable morphological and social environment for the generation and success of such activities, and that, conversely, insurgency (in one form or another) has therefore proved to be an important function of cities – shaping urban life, morphology and planning, both now and in the past.

TYPES AND STAGES OF INSURGENCY

A spectrum of increasing lawlessness and seriousness of the challenge to the public order of the city (and the authority of those charged to maintain it) can be envisaged. This could begin with crime, and no society (even More's Utopia) is free of some sporadic intentional disregard for its regulations. The difficulty of eradicating crime often resulted in its containment into specified areas of the city where the municipal writ no longer ran. The 'rookeries' of nineteenth-century London were, in modern terminology, 'no-go' areas and were by no means exceptional among large European cities. A growing disregard for public order can escalate to the intensity of an urban riot, a more or less spontaneous and loosely controlled challenge on the streets to the government's ability to maintain public order. A similar challenge to that same authority can be mounted by urban terrorism. This is an organized and purposeful attempt to subvert public order by actions designed to undermine public confidence in the ability of the authorities to provide internal security, and to induce an atmosphere of panic through random attacks upon certain chosen categories of target. A distinction between urban terrorism and urban guerrilla operations is that the latter is a direct confrontation with the official security forces, rather than an indirect challenge through terrorist attacks on 'soft' targets. With guerrilla actions the boundary between disorder and warfare has been crossed, although the conventional forces are attacked by non-conventional means and by irregular forces rather than by recognizable military units under the control of a recognized legal authority. When such guerrilla forces are in sufficient strength to confront openly and permanently their conventional opponents, then a state of open insurgency has been reached within the city – which has become a battleground for the opposing forces.

Such a spectrum of urban disorder contains the suggestion of escalation both in the seriousness of the challenge to the established urban government and severity in the likely response of the security forces. There is, of course, no inevitable progression from street vandalism to open warfare. Equally, such a classification is based on the level of violence and not on its motivation – if only because, although personal economic motives are more likely to be found at the lower levels and wider political motives at the higher, there is a considerable overlap between the two. 'Banditry has a long tradition as a form of political protest' (Niezing 1974), and urban terrorist and guerrilla operations undertaken for political ends are frequently financed by robbery and extortion. Urban riots, such as those of the American black ghettos of the mid 1960s (see Salert and Sprague 1980; Targ 1983) or the disturbances in a number of British cities in 1981 and 1983 (Peach 1985) may be motivated by a mixture of long-term political demands and short-term economic opportunism, as well as on occasion being orchestrated as a cover and diversion for terrorist operations – as was openly advocated and practised as the 'foco' technique (Miranda 1988) especially in Venezuelan cities between 1961 and 1964.

Nevertheless, despite the overlapping of motives and the difficulties of attaching a precise label to an often confused situation, it is helpful to make some categorization of the challenges to urban order, and to maintain some precision in the use of such terminology, if only because the way cities are used, and the sort of response of counter-insurgency forces, is quite different in different situations.

> The escalatory model serves the same role for insurgency as Clausewitz's notion of absolute war did for coventional conflicts: it is an ideal that is never quite reached. When examining an insurgency we should bear in mind the model – then look for the circumstances that make it different.
>
> (L. Ashworth 1990: 44)

The above spectrum of mounting levels of violence and of seriousness of the threat to established authority needs some underpinning in theories of revolutionary change, if the description of the results in the city are to be explained in terms of some logical processes. Although there are more theories of revolution than revolutions, and more theorists contributing to the copious literature than activists practising their craft, there is no single, clear-cut, accepted, general model of the causes, course or effects of urban insurgency. There is, however, a broad measure of agreement, that could be stretched so far as to be designated a developmental model, which recognizes a number of distinct phases of insurgency, in a progressive continuum (as argued by Moss 1971). Although the terminology varies with the writer, and the timing of the phases with the case study from which they are derived, it is usually possible to recognize four such stages.

First, there is a period of *preparation and propaganda* in which the preconditions for insurgency are established (these include the creation of a revolutionary organization and the recruitment of its personnel). These must then be equipped with the means of offering violent challenge to the authorities, which presumably implies the indulgence in a number of criminal activities – if only to acquire weapons, money and intelligence from home and abroad. Political activities in this phase include the establishment of objectives, the identification of the social or cultural groups from which support is to be sought, and the identification, creation and publication of grievances likely to gain that active support or passive sympathy.

This preparation is then utilized in the following stage – best described as '*urban terrorism*' – although this expression is often loosely applied as a pejorative term by the authorities to many different types of illegal activity.

Terrorism should be more narrowly defined as organized and directed acts of violence designed expressly to undermine confidence in the capacity of the authorities to maintain order, with the twin purposes of demoralizing the particular groups to which more or less random violence is offered – such as the security forces or civilian groups loyal to, or working in support of, the authorities – while simultaneously demonstrating to the potentially supportive social groups both the effectiveness of the insurgents and the impotence of the authorities. In order to be able to proceed to the next phase, that of guerrilla operations, the insurgents need the active help of a portion of the civil population and at least the passive neutrality of many more. The purpose of the terrorist phase is to obtain that support and sympathy. This can be achieved in two ways. First, the security forces can be provoked into a reaction that initiates an ascending spiral of violence in which the insurgents can assume the role of protectors of particular social groups against security forces whose responses fail to distinguish terrorist from non-terrorist. Second, terrorism can be used against the population supporting the government and providing the personnel for its services, thereby demonstrating the powerlessness of the authorities to protect this group, whose allegiance is thereby weakened. It is worth remembering that, as in most such insurgencies, the majority of terrorist targets during the Mau-Mau uprising in Kenya in 1952–6 were Kikuyu, and during the EOKA campaign in Cyprus in 1955–9 were Greek Cypriot (Laqueur 1977a).

In either event, terrorism has distinct objectives – being in theory neither mindless nor random in its use of violence – and is only the creation of the condition necessary for proceeding to full-scale guerrilla operations. In practice, however, such violence once practised or provoked is difficult to control or predict, in terms of: choice of targets; recruitment of insurgents, some of whom may have criminal rather than political motives; and popular reactions. A smooth transition to the next development stage is often difficult or, for most terrorist groups, in fact never occurs. This has resulted in many terrorist movements – such as the Red Army Faction in West Germany, the Angry Brigade in Britain, the Weathermen in the United States or the Front Liberation Quebec in Canada – being only small groups of disaffected eccentrics engaged in what are seen as no more than criminal activities, which can be successfully handled as such by the relevant police authorities.

Guerrilla war, as both words tautologically imply, is undoubtedly warfare, in which the military forces of the government are confronted by protagonists who see themselves, and organize themselves, as combatants. The difference with conventional warfare lies partly in the type of operations mounted (namely, hit-and-run activities rather than the attempt to capture and hold territory through formal battle lines) and partly in the fluidity of the guerrilla army whose personnel alternate between combatant and civilian status according to circumstance. In military terms, a guerrilla campaign combines tactical offence with strategic defence.

The evolution to the following and final phase of the revolution, that of *open insurgency* (indistinguishable from open warfare) may occur as a result of the success of the guerrilla operations, allowing the participants to become regular soldiers and take the strategic offensive in the final stage in the victory of the revolution which now assumes the reins of power.

All commentators stress that there is no inevitability in the escalation implied in these progressive stages, and as Mao argued, the course of revolutionary circumstances dictates a tactic of flexible oscillation backwards as well as forwards along the continuum (Mao Tse-tung 1963: 212). Victory was achieved by the insurgents in Rhodesia/Zimbabwe or South Arabia even though they were barely into the 'urban terrorism' stage. It is interesting, however, that not only is there a broad measure of agreement about the usefulness of such a model among revolutionary writings, but that the handbooks of counter-insurgency, intended as instruction manuals for the security forces, generally concur. The objective of counter-insurgency being, first, to recognize as rapidly as possible the stage of insurgency being practised, to institute the counter measures appropriate to that stage (chosen from the quite extensive armoury of such measures developed in the last 30 years), and thereby prevent an escalation along the continuum to the next more serious confrontation, and thereafter force the insurgents back into earlier phases.

WHY CITIES?

The insurgent activities discussed above can take place in many different geographical environments, and examples can be found of such activities in jungles, forests, mountains, swamps, cultivated lowlands and most other milieux. They are by no means confined to cities. This ubiquity would in itself be no argument for underestimating the importance of insurgency in shaping the urban scene or for ignoring the important influence of the characteristic features of cities upon the origins, course and impacts of such insurgency. However, it is the contention here that insurgency is relatively more important in urban than rural areas, if only because the population densities of cities allow it to be more apparent. In addition, aspects of the urban environment render it particularly suitable for many forms of insurgency and revolutionary lawlessness, to the extent that the first two phases of the revolutionary process outlined above (namely, preparation/propaganda and terrorism) are invariably urban, even when the latter two (guerrilla and mobile warfare) take place in rural areas.

Two sorts of argument can be advanced which, to an extent, overlap. The first is that cities cause insurgency, in the sense that many of the possible causes of revolt are likely to be present in greater quantity, intensity or at higher concentrations than in rural areas. The second is that cities provide a higher density of opportunities for revolt, or for the successful maintenance of insurgency once begun, regardless of the character of the original causes. A problem in sustaining the first argument is the absence of agreement among political scientists about the causes of confrontation with established authority (see, for example, Rule 1989), whether such confrontation is relatively minor individual law-breaking, collective riot, or full-scale revolutionary uprising. Whatever the causes, however, and whatever the set of psychological or social conditions that nurture the professional revolutionary leader – or the potential active recruits to the revolutionary struggle (see Moss 1972) – cities will, by definition, possess more of them to the square kilometre. Insurgency is a minority pursuit, especially in its early phases, and participants may only be present in the necessary critical mass in large cities, which become the only places where revolution may be possible.

In terms of opportunity, cities have two important characteristics. They are the concentration points of the established wealth and power whose seizure is the ultimate objective of the insurgency, and they also contain the concentrations of transport, communications, production and administration that support that wealth and power. Cities, and especially large capital cities, are the self-conscious show-cases of the country and thus of the governing regime. The city presents a range of targets distinguished not only by their physical density but also by their practical and symbolic importance. Attacks on, or disturbances around, such targets cannot be met with indifference, which for the insurgent is the worst possible reaction.

But if the city is more vulnerable to attack, it is not only potential insurgents who are aware of this vulnerability, and who can take advantage of the concentration of potential targets. The same consciousness will result in it being more heavily, and because of the very physical concentration, in many ways also more easily, defended by the security forces – whose power will also be concentrated in the urban centres, especially the capital cities.

The second characteristic of cities, it is argued, is that they provide a physical environment that favours insurgent operations by allowing them to capitalize on their advantages of flexibility, short-distance rapid mobility through different sorts of terrain, and the possibility of physical concealment (merging with the civilian population and the like), while minimizing their inherent disadvantages of lack of fire-power, inability to deploy large units, and lack of long distance mobility.

> Vietnam's jungles have no elevators or stairwells in their treetops but city buildings do – and a multitude of vacant rooms in which to flee. No jungle tree branches are as secure. The degree of security for city guerrillas is almost too imposing to suggest.
>
> (Rigg 1968)

Such an analogy between the physical characteristics of the tropical jungle and the morphological patterning of cities can be regarded as far-fetched, but the point that both environments share attributes advantageous to insurgent operations, and conversely difficult for conventional forces, is validly made and well known (Ney 1958).

The argument about the 'correct' place for revolution, especially the suitability of urban as opposed to rural environments, has been long and bitterly fought, becoming a point of ideological faith among many revolutionary theorists. What might be termed 'classical' Marxist theory was centred around the critical role of the urban proletariat, and the countryside and its inhabitants were relegated to being a reactionary irrelevance. A dilemma that soon became apparent in the Russian case, however, was that although the urban proletariat could initiate the revolution, armies composed mainly of peasants were needed to defend it (Ellis 1973).

The Paris Commune of 1871 was elevated to a special significance in the hagiography of the revolutionary left, and its defeat therefore had a profound effect. To many it seemed as if the balance of advantage had tipped decisively in favour of the security forces, who could mobilize their firepower (including the first effective automatic firearms) to advantage in the broad boulevards of the modern city (Jellinek 1971). To many, this signalled the necessity to retreat to the countryside, where distance and terrain inaccessible to the wheeled vehicles of an increasingly sophisticated and thus supply-dependent military, offered refuge.

The practical examples of Lawrence's 1916 Arab revolt, the Moros and (later) Huks of the Philippines, the rural guerrilla in Mexico, and many others, appeared to offer the best model for insurgency. The essential failure of the rural anarchists, such as Zapata in Mexico or Makhno in the Ukraine, to dominate the urban centres of political and economic power was overlooked. The most influential writings were undoubtedly those of Mao Tse-tung, based on the authority of what remains the world's largest successful revolution. Here the importance of rural mobility, rural guerrilla bases and the support of the rural population was stressed, with the cities being isolated like islands in a sea of rural revolution, succumbing inevitably and consequentially. This contrasted not only with the Russian example but also with the Chinese nationalist insurgency experience of a generation earlier, both of which were urban. Maoist strategy was thus a response to previous urban failure, in which the Canton rising of 1927 loomed large (Ellis 1973). The subsequent influence of Maoist doctrine is to be found in direct applications not only in the Indo-China of Ho and Giap, and the Cuba of Castro (where the victories of the Sierra Madre overshadowed the contribution of the operations in Havana), but was also felt (if not always consciously acknowledged) in such diverse rural insurrections as those of Malaya (1948–60), Algeria (1954–62) and even Kenya's Mau Mau (1955–9). The idea that guerrilla war was particularly suited to peasant societies rested upon three technical conditions, namely: inaccessible terrain; a dispersed and thus difficult to police, population; and a society whose horizons were limited to local conditions. Such societies, it was argued, would be especially responsive to agitation (Ellis 1973). The assumptions of Clausewitz (1976) that insurrection demanded rural inaccessibilty, and the strictures of Castro that 'the city is the graveyard of revolution', coincided.

The causes of the reaction to this belief in the countryside as the proper scene of revolution, and the equally dogmatic swing back to the cities, evident by the 1970s, are difficult to disentangle. Even self-consciously rural revolutionary movements recognized that the final death blows to the government should be meted out in cities: the problem of timing this transition can be demonstrated by a string of failures: such as the original Sandanista rising in Nicaragua in 1927–33; the 'shining path' attacks on Lima; and even the disastrous (in military, if not diplomatic, terms) 1968 Tet offensive in Vietnam. Disillusion as a result of a series of rural failures was most strikingly personified by the débâcle in the jungles of Bolivia and the death of Guevara in 1967. Security forces were becoming more efficient at adapting to rural conditions – such as in the use of the helicopter, which largely negated the guerrilla's advantages on rough terrain by its vertical as well as lateral mobility. Urban operations, which in the traditional view had been seen as part of the final stage of a mass uprising, were increasingly used for furthering the first stage of propaganda and recruitment (Niezing 1974). By 1971 there were more than thirty urban insurrectionary groups operating, encouraged by the new model of success, Montevideo's Tupamaros. The Sandinistas in Nicaragua, the Euzkadi in the Basque country, the Provisional IRA in Northern Ireland, and the FLQ in Montreal were all part of the new orthodoxy, in which: 'increasing urbanization is making irrelevant nonsense of much of the detail of the revolutionary war theories of Mao Tse-tung and Che Guevara' (Burton 1975: 11).

The new urban orthodoxy of the 1970s had as little practical success as the preceding rural one, and urban 'foco' techniques frequently degenerated into smaller élites (see Table 4.1), easily criminalized and treated as such (Mack 1974). Historical experience therefore provides a range of answers to the urban/rural question, the complexity of which is apparent in the selection of examples in Table 4.2.

URBAN CRIME AND LAWLESSNESS

Urban crime is the one aspect of this continuum of confrontation with established order and its rules which has consistently attracted the attention

Table 4.1 A classification of insurgent appeal and environment

	Appeal	
Environment	*Elite*	*Mass*
Urban	Tupamaros	Paris Commune
Rural	Guevara (Bolivia)	'Peasant revolts'

Source: Adapted from Niezing (1974)

Table 4.2 Choice of rural/urban environments in some cases of insurgency

	Personnel			*Terrain during phase*		
Case	*Leader-ship*	*Member-ship*	*Ideology*	*Terrorism*	*Guerrilla operations*	*Open insurgency*
China 1926–49 (PRC)	U/R	R	R	—	R	R(U)
Malaya 1948–60 (PCAJA)	U	U/R	U	R	R	—
Cuba 1954–9 (Fidelistas)	U	R	U	U	R	U/R
Cyprus 1954–9 (EOKA)	U	U/R	U	U/R	R	—
Algeria 1954–62 (FLN)	U	R(U)	U?	U/R	R	R(U)
Vietnam 1958–72 (NLF, Vietcong)	U	R	U/R	U/R	R/(U)	R/(U)
Uruguay (Tupamaros)	U	U	U	U	—	—
Cambodia (Khmer Rouge)	U	R	R	R	R	R

Notes: U = dominantly urban; R = dominantly rural; / = and; ( ) = additionally; — = not applicable
Source: Adapted from Laqueur (1977b)

of spatial scientists and spatial planners. There have been many attempts to relate the incidence of law-breaking or law-breakers to social and morphological features of the city (see, for example: Georges 1978; Herbert 1982; Davidson 1981; Evans and Herbert 1989). It is not the place of this chapter to summarize this extensive body of work (into the place of crime within a social ecology of the city). Most crime is committed by individuals for a wide range of economic or socio-psychological reasons, and plays no part in attempts to subvert the established order in cities. However, some specific sorts of crime, together with those antisocial activities which may or may not be treated as crime (such as vandalism, graffiti daubing, littering and even noisy or boisterous behaviour), do play various roles in the process of insurgency. This leads in consequence to defensive reactions on the part of those responsible for public security, and by individual citizens concerned for their personal safety. The authorities react with situational crime prevention as part of the armoury of urban defence, and individuals fashion their behaviour according to an 'urban geography of fear'.

As with political revolution and large-scale public order disturbances, much less has been written on the detail of the role and significance of the urban environment as such. Most commentators reflecting on the spatial patterns of reported criminal offences, have concluded that cities and (especially) their inner areas, should be regarded not so much as hatcheries for criminals, but more as peaks in a crime opportunity surface. This results from the spatial concentration of targets of opportunity, whether people or property, as well as the concentration of population from which criminals are drawn. Each district of the city will thus have its own pathology of crime related to the opportunities it offers, with the micro-features of the urban morphology contributing to a landscape that favours or deters such activities. If 'offences have their own ecologies of planning, neighbourhood layout and design' (Harries 1974), then offences against public order and street crimes against people and property will be related principally to the opportunities of the inner city. In terms of the physical environment, such activities will be favoured by a set of environmental conditions similar to those favoured by guerrillas (namely, possibilities for concealment, ambush, and rapid retreat through terrain less accessible to the surveillance or pursuit of police). An irony of many modern cities is that such a landscape has frequently been created by modern architectural and planning conventions, which have produced the enclosed and indefensible public spaces, hallways, lifts and underpasses of many housing projects and simultaneously an inaccessibility to motorized policing of pedestrian city centres. Newman's (1974) concept of 'defensible space' and implied criticism of the 'defensibility' of much post-war urban development was widely adopted by planners, and became a conventional wisdom in much detailed urban planning – see, for example, the Dutch national planning report on crime and design (RPD 1985).

The relationship between crime and the more serious challenges to authority can be traced through the law-breaking acts of politically motivated insurgents, which are frequently indistinguishable from similar offences committed for individual economic gain; through the engendering of an atmosphere of insecurity among citizens with regard to public order, especially fears for the safety of their persons and property. The existence of what Tuan (1979) has called a 'landscape of fear' can become something of a self-fulfilling prophecy, in which the belief that places are unsafe both deters the deterrent presence of the law-abiding and attracts potential law-breakers. Mawby (1984), among others, has examined in detail this question of the perception of the safety of areas. Fear may lead in turn to an undermining of public confidence in the efficiency and determination of the security forces and of the governments operating them, and through reactions in both official policing policy and the property market to the creation of a defensive urban morphology.

The walls of the European medieval city imposed a physical control on the movements of those within, and upon the entry of undesirables from without. The 'watch' that manned such walls and their gates had police as well as military functions. Most cities reinforced these measures with street curfews enforced by patrols, and numerous regulations limiting the bearing of arms in public places to a state monopoly. The Chinese city during the Han and Tang periods (i.e. up to around AD 700) supplemented the almost universal external wall with a compartmentalization of the city by means of walled wards, within which stood the individual houses, each of which was itself walled. All three sets of gates (i.e. those of house, ward and city) were closed and guarded at night, providing (at least in theory) an absolute control on movement and thus on public order (Tuan 1979).

Such physical defence systems within the city may be as much a sign of the absence of effective public order as of the presence of an overzealous authority. The lack of order within many medieval European cities – whether stemming from individual crimes of opportunity, or (as was frequently the case) organized feuding and factionalism between families, clans and political parties – led to similar defence reactions of fortification within the city. The mini-fortresses and their towers of the quattrocento towns of the Italian Romagna were a reflection of this need for individual provision of physical safety within a lawless city. Jerusalem in the nineteenth century was divided into a series of self-policed protective compounds into which different ethnic and religious factions withdrew at night (Cohen 1977). It is worth remembering that as late as the middle of the eighteenth century, in a country such as Britain (which had proceeded further than most in developing concepts of political responsibility and ordered government), a journey through the city streets generally necessitated the private employment of both street lighting and physical protection by 'linkmen'. In our times, in many North American cities, the fear of street crime and the lack of confidence in the policing authorities' provision of protection has encouraged private initiatives to provide at least 'islands of security' within the city – which are not far removed from those of fifteenth-century Sienna or Bologna.

The idea of 'the neighbourhood as fortress' (Davidson 1981) is thus not new, nor is the paradox that the more security is sought by physical protection against the city outside, the less powerful are the social controls upon that city, and thus the more unsafe it becomes.

Harries (1974) has graphically described what he terms the 'model defensive environment', which is in part a 'worst case' prediction and in part not far from a description of reality in a number of cities in the United States or in parts of cities (such as Belfast – discussed at length later) where special threats to personal safety exist. According to Harries, cities would consist of five sorts of area, distinguishable in terms of their security characteristics (see Figure 4.1). The *central business district*, protected during business hours by the presence of people and a visible police deterrence, would be sealed off at night and policed by electronic surveillance, as a modern equivalent to the medieval curfew and street watch. The *residential districts* of the inner city would exist within fortified compounds, with movement into them controlled by guarded gateways; security would thus be purchased as one characteristic of the house and neighbourhood, with residential choice being determined in part by how much of such security is required and can be afforded. The *outer suburbs* would be protected largely by distance from

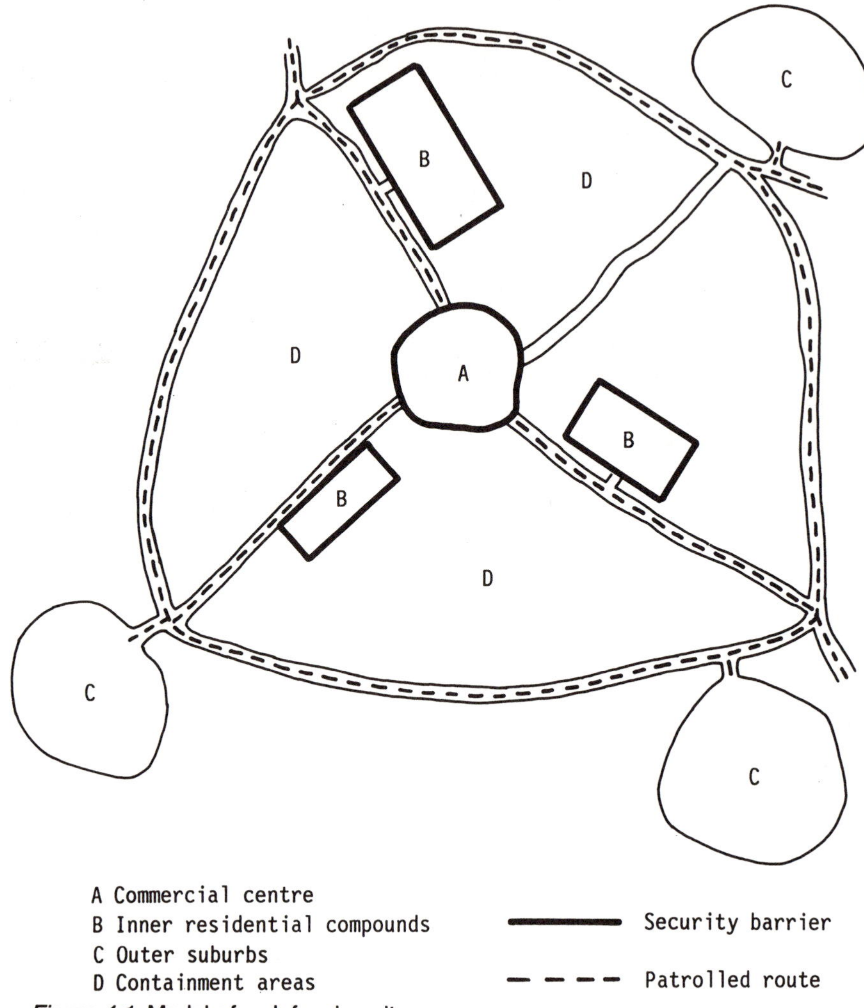

Figure 4.1 Model of a defensive city

each other and from the inner city, as well as by the homogeneity of their populations. Such suburbs would be self-contained in many commercial and public services in order to reduce the need to visit the inner city. The various *districts of the city* would be linked by 'sanitized corridors' which would connect 'safe' areas to each other by policed high-speed highways, to be traversed in locked cars. In the *rest of the city* crime would be contained if not controlled, again a policing pattern familiar in the nineteenth-century, high-crime areas of most European cities.

URBAN RIOT

Collective breaches of public order through riot, unrest and public demonstration need not be specifically urban, although even the Peasants' Revolt of 1381 culminated in a march on London. The very concentrations of population makes an urban setting more likely, and the symbolic importance of many urban places, as well as the proximity of governments, provides more opportunities. Nor are these public disturbances particularly a modern phenomenon, although modern communication techniques may render them more effective as political demonstrations, and spread the news of such occurrences more rapidly through potential rioters, creating a 'copy-cat' sequence. Urban riots were a fairly regular and expected feature of the urban scene, so that most cities had their recognized congregating points for the staging of such demonstrations, which survive in such spaces as London's Trafalgar Square or The Hague's 'Malieveld'.

The definitive line between legal and illegal demonstration is difficult to draw, and is as much a matter of prevailing custom and compromise as law. So also is the distinction between crime (especially street crime) and riot, and many of the defensive techniques described earlier stemmed as much from the perennial fear of the mob, that survived well into the nineteenth century in almost all European cities, as from fear of individual criminal acts.

Similarly, the relationship between urban riots and urban terrorism and guerrilla insurgency is complex. Many collective disturbances have no political implications. Jarowitz (1969) distinguished 'communal' from 'commodity' riots on the basis not of the motive but of the target, the first being an ethnic or social group and the second being property. To these could be added general hooliganism – associated in the 1980s in western Europe particularly with football matches – in which the targets appear to be randomly chosen. Disturbances may even, within certain understood limits, be tolerated. For example, many societies, both now and in the past, have specified times of acceptable 'misrule' or 'carnival'. However, many riots may be more-or-less spontaneous and undirected outbursts of anger and frustration at real or imagined grievances.

In what was termed the 'long hot summer' in the middle of the 1960s, when the residents of the black ghettoes of American cities turned to widespread looting and burning, the underlying grievances might have had a logic in the years of discrimination or neglect, but the timing and choice of cities appeared to be unpredictable and disorganized (Button 1978). Similarly, Georges' (1978) study of the Newark riots of 1975 found it difficult to relate either the particular timing or location to specific grievances. In a detailed analysis of what Lewis (1976) argues is an unbroken, 200-year tradition of violence in the United States, he concludes that the most important locational characteristic is simply the population size of cities. During the period of greatest media interest (1964–8), 80 per cent of all property damage and 60 per cent of all injuries occurred in just five cities (Los Angeles, Newark, Detroit, Washington and Chicago). The far less severe but in some respects similar disturbances in British cities in 1981 and 1983 again proved difficult to predict or explain on grounds of the social, economic or size characteristics of the cities concerned (Peach 1985).

Mass demonstrations organized for political ends, whose objective is to confront the authorities by acts of collective law-breaking, have in recent years become an accepted, legitimate part of the political scene, even in parliamentary democracies. Indeed, the experience of the *annus mirabilis* of 1989 in eastern and central Europe suggests that they may have become a necessary preliminary to the establishment of such a democracy. In addition, mass urban rioting can form part of a more extensive range of insurgency activities: for example, as a cover for terrorist or guerrilla acts of assassination, sabotage or subversion of the security forces, or as part of a propaganda and preparation phase of full-scale revolution – as grievances are made public, the security forces provoked into an over-reaction that increases such grievances, and potential activists for the later stages of insurgency are recruited. The large scale 1986/7 disturbances in the black townships of South African cities, and the Palestinian 'uprising' since 1987 (*Intifada*) in the towns of the West Bank and Gaza Strip (although censorship in both cases made it difficult to assess) appear to be examples of such a mixture of popular spontaneity and political orchestration, whose purpose is to consolidate the legitimacy of an insurgent organization both internally and externally. Such consolidation may occur as a result of the actions of security forces in response to an existing riot. In 1990 up to 20,000 Soviet troops had to force entry into the central area of Baku, which had already been barricaded by insurgents, in order to secure government and party buildings and the Subunchinsky railway station. The resulting casualties did much to legitimize Azerbaijan separatism.

The reactions of the authorities to riot, or in the longer term to the fear of the possibility of riot, has throughout much of urban history been either to fortify refuges within the city (the 'citadel solution') or to remove themselves from the city, or from the parts of it where the actions of the mob were most prevalent (the 'Versailles solution'). The former option was adopted by the Norman conquerors of eleventh-century England, where the castle on its motte was designed to protect the new government from the existing citizens and to cow the city into submission rather than protect it from external attack. Such a use of the castle was as much the rule as the exception through much of the Middle Ages. London's Tower, Paris's Bastille, Utrecht's Vredenburg, and many other such citadels, sheltered the urban government against enraged citizens whose lawlessness was allowed to burn itself out in the town outside.

The vulnerability of national, or even in the case of Rome and Constantinople imperial, governments to the urban mob of the capital – whose demands had to be placated at the expense of national policies – was an argument encouraging a move of courts and government apparatus beyond its physical reach into the more peaceful countryside outside the city. Bourbon Versailles of the seventeenth century was only a later and larger-scale version of the English Westminster option of 600 years earlier. In its extreme form, this option was a not unimportant consideration in the creation of new government cities established beyond the disruptive influence of the metropolis. The Hague was a haven away from the tumult of the Holland cities, whose mobs had already murdered the government of de Witt; Washington was removed from Philadelphia and Boston whose unruly citizens had already led a successful revolution; and, in more modern times, the foundation of cities such as Brasilia and Islamabad have more in common with seventeenth-century Versailles then the spacious, imposing architecture alone.

At the more micro-scale it has long been clear that there is a relationship between the details of the street and building block-patterns, and the effective deployment of the military resources of the security forces, although this relationship can be oversimplified. It is true that narrow, irregular streets and spaces, deny to the security forces a number of their inherent advantages – especially their capacity to manoeuvre and deploy disciplined units, and to bring their superior fire power to bear at crucial points. Such an urban morphology is typical 'close country' in military terms, making the use of mobile forces (whether cavalry, tanks or armoured personnel-carriers) less effective. Fields of fire are seriously truncated by buildings reducing the effectiveness of artillery, and increasing the reliance on short-range, close-order small arms, where the citizens comparative disadvantage is likely to be least.

Many of these disadvantages, however, apply also to the rioters – who equally have a need for open space and broad lines of movement, if they are to exploit their weight of numbers. In addition, the psychology of the mob requires the visible presence of a critical mass of participants in order to create the atmosphere of an unstoppable tide of protest which will overwhelm opposition. Beijing's Tianamen Square may have been an ideal, open and symbolic space for a mass challenge to the established order but it was equally ideal in 1989 for clearance by the military forces of that order.

Standard police tactics – known as the 'herding sheep technique' according to Methuin (1970) – are, in order of priorities: first, to cordon off vulnerable areas; second, to contain the rioting within specific areas of the city; and, third, to enter the riot area and disperse the rioters. The detailed urban morphology plays an important role in all three stages. Bodies of rioters and demonstrators are broken up by forcing them away from the open spaces of squares and boulevards and channelling them into narrower streets and alleys. This may be undertaken either by the physical pushing by large numbers of unarmed police, as in Britain (as described in detail by Deanne-Drummond (1975: 108), or by resort to anti-personnel projectiles.

In both cases it is important that the street pattern not only allows escape but is precisely the sort of 'close country' whose morphology itself splits rioters into smaller more manageable groups whose impetus will be lost by dispersion. Many of the occasions when such tactics fail, and result in either a collapse of the security forces or an over-reacting 'police riot' can be related to failures to use the morphological pattern correctly: for example, by herding rioters into an area from which they cannot escape – 'the bursting boiler' situation (Methuin 1970).

The textbook example of the redesigning of a city in order to render it more defensible against urban riot, is the work of Baron Haussmann, prefect of Paris during the reign of Louis Napoleon, who ironically had himself come to power with the help of the Parisian mob, whose importance he therefore appreciated. The barricades had been erected across the streets of Paris no less than eight times between 1827 and 1849. The new broad boulevards, such as Rue de Rivoli, Boulevard Sevastopol and Boulevard Voltaire, that were driven through some of the remaining close-packed, working-class quarters of the city intra-muros dramatically improved access and fields of fire. More of the city could now be brought under control by 'a whiff of grapeshot', as the first Napoleon had demonstrated in 1794. Such boulevards into the heart of the city were often linked to new barracks (such as the Caserne Vitrine on the Rue de la Republique) so as to facilitate troop deployment into the working-class districts. The double irony was that, first, these precautions, intended to frustrate an urban uprising, were taken immediately before the most serious such revolt in a major capital city up to that time (the Commune of 1871), and, second, that the demolition and rehousing needed to create these precautions may well have contributed to the sense of grievance which fuelled the very uprising they were designed to prevent.

The town-planning ideas put into practice by Haussmann in Paris had numerous imitators in French provincial cities in the third quarter of the nineteenth century (Sutcliffe 1970). They were fashionable notions of how a 'modern' city should be designed rather than self-conscious attempts to create a riot-proof urban morphology, and any security advantages must have been welcomed as a fortuitous side-effect. It is, in any event, too easy to link such developments with an individual city and an individual initiator. As Hall has traced in detail for a number of European capital cities, an important motive for rebuilding the city after 1850 was 'removing the environments which encouraged political disorder' (Hall 1986: 28). A generation earlier, the rebuilt ring-boulevards of Hapsburg Vienna had facilitated suppression of the street disturbances of 1848–9 (through a combination of regular artillery and cavalry) after they had come close to the overthrow of an imperial dynasty. The first proposals for the Vienna 'ring' zone was that it be left open as a 'cordone sanitaire' between the governing and the governed classes. The plan eventually adopted included some new barracks, notably the Franz Josef Kaserne, for the housing of a 'rapid-reaction force' for use in the working-class suburbs. Similar developments of a new road system, together with barracks for the troops who could deploy along it, was found in cities as diverse as Barcelona and Stockholm (Hall 1986).

A hundred years earlier still, the baroque *residenzstadte* of a host of German ruling kings, princes, dukes and bishops had restructured their medieval capitals with broad processional ways, central squares and rotundas, and a rigid, geometrical patterning of streets (Mumford 1961). It would be far-fetched to suggest that such efforts sprang entirely, or even chiefly, from the rulers' fear of the unstable mob that infested the medieval streets and alleys of the old town. A desire for architectural display and for the public manifestation of the symbols of power were stronger motives than the creation of fields of fire for artillery or space for the drilled manoeuvring of bodies of regular household troops. Indeed, it is difficult even to associate these planning forms with a political totalitariansm that invites popular insurgency when planned capital cities from Washington to Canberra have tended to adopt such styles as being appropriate to government.

It is also salutary to remember, as a counterpoise to an over-deterministic view of the relationship of urban design and successful riot control, that some of Europe's most riot-prone cities possess the very design characteristics that favour such control. Barcelona, the city whose reputation for successive political rioting in the early decades of this century was such that the city authorities were rumoured to have numbered the paving stones so that they could be easily replaced after each misuse as barricade-building materials, had been rebuilt to the broad-grid plan of Cerda. Similarly, an impartial insurgent seeking a suitable urban layout for mass demonstrations and confrontation against conventional forces would have placed St Petersburg very low on a list of preferred cities. Not only is it dominated by broad boulevards but also its river crossings are easily controlled by a limited number of bridges, some of which were actually movable. The fact that successive governments were overthrown with the help of precisely such mass insurgency in 1917, but successfully resisted it in the same city in 1905 and 1920, merely demonstrates that the spatial patterning is only a contributory factor – of secondary importance when compared with the level of skill and resolution of the security forces and those controlling them.

URBAN TERRORISM

The failure of the armed uprising of the Paris Commune of 1871 when confronted by regular forces, marked, according to many commentators (see Burton 1975), a turning point in the history of urban insurgency. What was obvious to many was the apparent helplessness of untrained, poorly armed, loosely disciplined citizenry commanded by amateur leaders in the face of a combination of new military technology (including an accurate infantry rifle, the first quick-firing automatic anti-personnel weapons, and mobile field artillery), a resolute national government able (thanks to new transport technology) to mobilize the country against the capital, and a new design of cities that appeared to have rendered the barricade obsolete. For three-quarters of a century the romantic myth, of revolutionary fervour inspiring the righteous masses in open insurrection against a timid reactionary government and its hireling soldiers, that had been launched in 1789, appeared to be an instant recipe for success, proven in Brussels and Paris in 1830, and in most of the capitals of Central Europe in that year of revolutions, 1848. This era came to an abrupt end in 1871 and the alternative strategy of urban terrorism was born.

Such an interpretation of the historical development of insurgency exaggerates the importance of a single event in one particular country, and underestimates the previous use of terror as an insurrectionary weapon. Examples of the deliberate use of attacks on 'soft' targets by small revolutionary groups designed to demoralize conventional forces, or control the allegiance of particular social groups, can be found in most historical periods (see Mickolus 1983 for a historical inventory of such groups).

However, such an interpretation of events towards the end of the nineteenth century does serve to underline some of the essential features of terrorism. It is by its nature a reaction of the weak against the strong, allowing small (often extremely small), poorly-equipped groups to create an impact out of all proportion to their size, resources, or degree of popular support. In this sense it has more in common with crime than with mass demonstration and riot. On the other hand, it demands a high degree of motivation and organization and a 'professionalisation of the art of insurrection' (Burton 1975). The popular view of the wave of assassinations that swept Europe in the last years of the nineteenth and beginning of the twentieth centuries, as being instigated by the stereotyped bomb-throwing anarchist, does scant justice to anarchism as a political creed, but does encompass many of the essential features of terrorism as a controlled, directed strategy of the professional revolutionary too weak to pursue alternative methods.

There is also an element of truth in the idea of urban terrorism as a phenomenon created at the beginning, and coming to fruition during the course of, the twentieth century. The objective is to create a feeling of insecurity and panic among members of the targeted group, whether these be a particular group of citizens, members of the security forces or governing administration, or prime ministers and tsars, any of whom must feel they could be chosen at random for attack (Johnson 1962). The success in instilling this insecurity is dependent upon the efficiency and openness of communications within the society under attack. Clandestine terrorism is a contradiction in terms: 'the propaganda of the deed' needs publicity. As

Stohl (1983: 1) put it: 'Political terrorism is theatre. It is drama of the highest order'. The development of the communications media and associated publicity industry has thus made the modern city, especially in open societies, particularly vulnerable.

The vulnerability of the modern city in the other sense of its concentration of targets, 'the eye of the octopus' (Stokely Carmichael, quoted in Burton 1975), has already been commented upon. It is not only that the growing complexity of urban society has increased the disruptive effect of terrorist action, thus magnifying its power, but also that the show-case functions of cities, their edifices, monuments and ceremonies present targets of immense symbolic importance. Assaults upon the show-case are an obvious means of obtaining the publicity needed for the success of terrorist aims.

The increasing internationalization of communications in the global village has been parallelled by an internationalization of urban terrorism, with targets being selected for their international news value (see Livingstone 1978, for a list of major terrorist attacks in the 1960s and 1970s). The group thus 'terrorized' may appear to have little direct relevance to the particular insurgent struggle and thus the argument that terrorism is a preparatory phase in which the support of a population is mobilized for an escalation into guerrilla operations appears to have been breached (Gurr 1983). However, the publicity gained from terrorist acts may be intended not only to impress the potential support needed in the area of insurrection itself but also for a number of other purposes, which may include competition in violence between different factions among the insurgents themselves for leadership of the struggle. The necessarily loose organization of such groups has always encouraged such factionalism, such as that between various groups of the PLO, or between the INLA and various wings of the IRA (Targ 1983). The advantages of 'internationalizing' a domestic struggle will encourage attacks that make international news, bringing the situation to the attention of an international public. Indeed, just a realization that an insurgent group exists will determine that the selection of targets be as spectacularly international as possible – such as the Palestinian attacks on the Munich Olympic Village in 1972 – even if that realization is obtained at the cost of world-wide condemnation. This internationalization clearly leads to a concentration at the highest echelons of the urban hierarchy and to targets such as international airports, public ceremonies with international significance, and even individuals endowed with international star quality. It is not surprising, then, that since the early 1970s cities such as London, Paris, Rome or Athens have been subjected to numerous terrorist campaigns for causes over which the inhabitants of those cities have neither concern nor influence.

The options open to security authorities responsible for preventing the use of cities as mass political hostages in this way are extremely limited. Urban-terrorist activist cells can be extremely small, while their potential targets can be many and dispersed through the city. Three main types of strategy may be attempted, supposing that the option of merely ignoring such attacks and thereby denying them much of their publicity value is politically and practically impossible.

The first is that they can intensify protective security, increasing surveillance on potential targets in the city. Deane-Drummond has suggested that:

> future layouts of police stations, telephone exchanges and other public utility centres should be designed with reference to their vulnerability to bombing, occupation by squatters, riotous mobs or terrorist groups.
>
> (Deane-Drummond 1975: 107)

Such precautions form the most ubiquitous and visible impact of 20 years of sporadic terrorism on Northern Ireland towns. In addition, the freedom of movement and operation of potential terrorists can be restricted by defensive measures such as controlled cordons around vulnerable areas, searches of the general public and the like. One of the more extensive examples of this was in the central area of Belfast, including around 300 retail outlets and about 25 per cent of the city's total retail floor-space. The constant use of the area for bomb-planting (1,800 bomb attempts between 1970 and 1975) so as to maximise publicity, provoked the following reactions: first (1972), a ban on unattended parked cars (a usual *modus operandi*); second, the sealing of through routes (1976); and, third, the creation of a security cordon accessible only after surveillance and search. Such operations, however, are not only extremely expensive in terms of police manpower, but also ineffective in apprehending terrorists – offering only a partial protection to targets through prevention. In Belfast there was a decline both in bomb alarms and in visits to the area, therefore the security segment was successively dismantled between 1981 and 1984 (Brown 1985). An even more important disadvantage is that at best they cause irritating delays and inconvenience to the public at large (which may be acceptable when the security advantage is obvious to the individual inconvenienced, such as at airports), but at worst they erode the very freedoms of movement and association that (in a democratic society at least) is the object of the terrorist attack. The police are caught in a paradox that efficiency in protection against terrorist acts may help to achieve the goals for which such acts are perpetrated. A high-profile security presence, restrictions on the activities and movement of a registered and identifiable population, limits on publicity and press coverage, and summary search, arrest and internment will be extremely effective in limiting terrorist activities. However, the cost may be the alienation of a previously largely neutral population, an unsympathetic foreign reaction (thus increasing external support for the insurgents), and damage to the democratic fabric of the society being so defended.

The opposite approach – namely, a 'hearts and minds' campaign to capture or retain the allegiance of the population, or at least separate them from the terrorists – is only in part an activity of security forces, and depends as much on political initiatives at government level. Combinations of the two approaches are, in practice, rather difficult to achieve, and an uncertain oscillation between them is an almost certain recipe for failure in both (see Cable's 1987 discussion of such policies in Vietnam).

Two policy options designed to neutralize the terrorist rather than protect the targets are open. The first is a police rather than military operation, although on occasion executed by the latter. It is the gathering of intelligence through normal police methods, supplemented by infiltration and clandestine quasi-legal penetration of terrorist cells and their destruction. 'Search and snatch', impersonation and even selective counter-assassination can be extremely effective – hitting terrorist groups at one of their vulnerable points (their uncertain discipline and organization) – so long as the intelligence upon which such operations is based is sound. The success of the West German police against the RAF and the Italian police against the Red Brigade and its imitators are notable.

An alternative aggressive policy, which does not have the same intelligence requirements, is counter-terrorism, which quite simply answers the terrorist acts, with similar random attacks designed to intimidate the population by using fear of reprisal as a means of separating insurgents from their potential supportive civilians. Such acts of terrorism, conducted by the security forces as a systematic policy, were popularized by the French army 'Guerre Revolutionnaire' school on the basis of their colonial experience but have been committed at one time or another by the governments of most countries. Frequently, the targets are urban even if the original terrorist acts were not. The British use of the aerial bombing of settlements in Iraq and Somaliland in the inter-war period (Kennett 1982) was mild compared with German counter-terror policy in both the First (the destruction in Leuven and other Belgian cities in response to 'francs-tireurs') and Second (the destruction of Lidice and Remadour being the best known examples) World Wars. More recently, there is an irony in one of the principal victims of the latter conflict being the most obvious practitioner of counter-terror techniques: the Israeli bombing of cities in Lebanon, and even Tunisia, is a response to terrorist attacks elsewhere. Even more indirectly, there is the example of the United States' bombing of Libyan urban targets in response to the suspected support of that government for terrorist attacks by others elsewhere.

The objections to the use of counter-terror are not only moral indignation and international condemnation at breaches in conventional acceptable behaviour but also that such acts are probably counter-productive in anything but the short run because of the publicity and sympathy they engender for the insurgent cause (whether temporarily suppressed by fear or not) at home and abroad. Indeed, such actions are frequently no more than a symptom of the frustration of security forces in countering urban terrorism by other means. However, it is worth remembering that, although urban terrorism is difficult to prevent, its history would seem to demonstrate that it is eminently containable, if only at a cost in terms of (generally innocent) victims. Very few urban terrorist movements have succeeded in their expressed aims of evolving to the next stage of insurgency, and those that have proceeded to guerrilla operations have usually been aided by external factors, especially by regular military operations. Thus urban terrorism is generally not so much a threat to the structure of urban government and urban society as a chronic condition of cities (especially international capital cities), and, thus (like traffic accidents, environmental pollution and such like) a cost of modern urban life.

URBAN GUERRILLA OPERATIONS

An initial problem in considering the city as the scene of guerrilla operations is that a number of commentators would doubt the logical possibility of the existence of urban guerrilla warfare (Ney 1958). In part this is a reflection of the traditional belief, already discussed, that only rural areas offer the terrain requirements of guerrilla forces (see Clausewitz 1976). Blanqui's handbook on urban insurgency technique was, for example condemned by Lenin for failing to distinguish between insurrection and revolution. In part this is also a reaction to the pretensions of many urban terrorist groups that they are a guerrilla army when in reality they are no more than 'half-baked criminals perpetrating isolated acts of violence' (Ellis 1975). Certainly, Laqueur's (1977b) exhaustive chronology of 126 'notable guerrilla wars', of which only four were dominantly urban, would seem to support this view, although of course it is only in the modern period that an urbanized world has offered urban opportunities to potential guerrillas.

If the conditions necessary for guerrilla operations are examined, then the strength and weaknesses of the city as a scene of operations can be assessed. Successful guerrilla operations are dependent upon the existence of: (1) a terrain suitable for the practice of distinctive guerrilla tactics; (2) the support of sympathetic populations for supplies, recruitment and intelligence, or at least the absence of a hostile population that will supply intelligence to the security forces; (3) the development of an organizational structure to control such operations; and (4) the possession of a reasonably secure base area for refuge, rest and refit. Cities are at least at no particular disadvantage with the first three. As has been argued earlier, a terrain suitable for evasion, sabotage, ambush and concealment, and conversely unsuitable for large scale deployment and long-range weapons, exists.

Guerrilla war, however, is more than just a military tactic, which can equally well be practised by units of the regular forces that are organized or act irregularly – such as the SAS or 'Green Berets' (Keegan and Holmes

1975). For the insurgent 'to approach the subject of guerrilla war as a purely military doctrine is to court disaster' (Oatts 1949). Close contact with the local population is of critical, political (as well as military) significance. Cities' very density of population would seem to offer some advantages over rural milieux, in so far as such contact for recruitment and for the exercise of organizational control over activists is concerned. Indeed, such standard counter-insurgency strategies – designed to separate guerrillas from their local support, strategic hamlet or fortified village policies (McCuen 1966) – are almost impossible to implement in urban residential areas. The problem seems to lie principally with the last requirement, that of a secure base, which cities can rarely provide against even moderately competent security forces. 'Cities just cannot house groups large enough' (Laqueur 1977b: ix), therefore urban guerrilla operations are necessarily small-scale and short-lived. In Ellis's (1975) list of 130 urban guerrilla operations, only twenty formed part of ultimately successful insurgencies. One solution is to 'live separately but fight together', which is generally beyond the organizational skills of most guerrilla groups.

It is not surprising, therefore, that cities have often been the scene of small-scale guerrilla operations, in this respect indistinguishable from terrorism conducted by insurgents whose bases lie outside the city – such as the Fatah operations in Gaza in 1967. Thus, although cities are rarely the location of successful guerrilla activities, they have in the past quite often been the scene of terrorist operations conducted by and in the support of, rural-based guerrillas – such as the Huks in the Philippines (1950–2). Marighela (1969) saw a complementarity between urban-tactical and rural-strategic operations. The interaction between terrorist operations in Havana and guerrilla actions in the Sierra Maestra in Cuba (1956–9) was important in defeating the Batista regime. EOKA terrorists in Cyprus (1954–9) mounted such operations in Nicosia, although based principally for long in the Troodos Mountains. In Aden (1967) the real guerrilla campaign was in the Radfan Mountains, although terrorist operations in the town of Aden itself, conducted by FLOSY and other groups, provided at least a distraction and at best some demoralization of the security forces, as well as important world-wide publicity, although the attempt to establish an urban guerrilla base area in Crater was frustrated with relative ease.

The urban dimension in the long FLN insurgency in Algeria is equally ambiguous. Although terrorist attacks on urban targets were widely used by both FLN and 'pieds noirs' groups, the two serious attempts to organize guerrilla operations in the cities both failed. In 1957 the FLN deliberately attempted to utilize the narrow streets of the kasbah in Algiers as a fortress, to be defended by the Muslim inhabitants, after the expulsion of elements regarded as unreliable, with cached small arms. Against skilful and determined security forces, it succumbed relatively easily. The second 'battle of Algiers' was less well prepared by the insurgents (this time the European settlers), and was fought out in the much less favourable terrain of the broad boulevards and squares of the European city. The 'pieds noirs' could only have succeeded if the loyalty or enthusiasm of the French regular forces brought against them could be undermined, and this did not occur (Horne 1977).

OPEN INSURGENCY

The transition from guerrilla warfare to open insurgency is difficult to make, and a number of successful guerrilla campaigns have prematurely escalated into open conventional warfare, with consequent defeat by regular forces – such as the FLN attempt to seize the Algiers kasbah in 1957, or the Vietcong Tet offensive in 1968. 'Urban uprisings are easy to start but are hard to win' (Wheatcroft 1983: 75). Nevertheless, successful urban insurgency, if it is more than an ancillary theatre to military operations elsewhere, requires that such a transition be made: that irregular partisans become disciplined full-time units and the enemy confronted openly in battle. The historical fact that such open insurgency is rare is attributed to the inherent difficulty in proceeding through the stages of development of insurrection as postulated, and also perhaps to the success of security forces in preventing such a progression. There is also perhaps the centuries-long instinct among the military (discussed in Chapter 5), that cities make unsatisfactory battlefields, and regular war in the cities is thus to be avoided. There are, of course, many instances of urban areas becoming (by design or not) part of far wider battlefields, and these are considered in Chapter 3, but examples of insurgents openly confronting regular forces in the city streets are few, and successful cases fewer still. A reason is not hard to find: the only effective defence which insurgents have against regular forces are guerrilla tactics: if these tactics are abandoned, the only hope is that other factors will intervene – such as regular assistance being forthcoming from elsewhere; the security forces being particularly disorganized, demoralized or subverted; or the conflict has the character of a civil war where neither side could be described as regular.

The best known case-study, which has achieved a particular significance in the annals of both insurgency and counter-insurgency is the seizure of Paris in 1871 by the communards, and it subsequent defence against the regular forces of the government ordered to retake it. Although this episode included guerrilla tactics (especially in its later phases), and the insurgents were not accorded regular status and were largely semi-trained, poorly disciplined but highly motivated volunteers, it was open insurgency as the attempt was made to defend the territory of the city, and the provisional government structure that had been established, behind the lines of its walls. The detail of the course of the engagement, which Jellinek (1971) describes in detail, seems to provide evidence of many of the relations between morphological patterning and the effectiveness of regular forces. The relatively easy breach of the static defence line of the walls by infantry with artillery support, followed by advances along the 'Haussmannised' boulevards (where, again, artillery and aimed rifle fire could be used to effect), was held up by the defensive bastions of the 'internal fortresses'. These were generally clusters of substantial buildings linked by barricades (such as the Trocadero, the Pantheon and around the Pont d'Austerlitz) which were more often than not eventually taken by outflanking rather than direct assault. The 5-day battle rapidly became a series of isolated engagements in the largely medieval 'close country' of small alleys and tight-set buildings of Belleville or Butte Chaumont where the last stands were made. It should be remembered that these areas were socially, as well as morphologically, favourable to the insurgents, being largely working-class residential districts and the 'home' areas to many.

The barricade, an improvized defence work intended to physically block an avenue of advance while providing shelter for defenders, has a special symbolic significance for armed revolutionaries. First named in Paris in the fifteenth century, it was decisively demonstrated to be obsolescent in the same city in 1871. Or, to be more accurate, its function as a defensive breastwork was shattered by a combination of spaces too wide to be quickly blocked and fields of fire allowing disciplined infantry to sweep away defenders, and artillery to do the same to both defenders and barricade with little difficulty.

The barricade was rediscovered by insurgents in Budapest in 1956. Its new function was to be an unexpected obstacle, erected in ambush positions with restricted fields of fire, intended to stop wheeled or tracked vehicles and thus create a killing ground enfiladed from the flanks rather than manned from the front – see the detailed instructions about how and where to build a barricade against armour in International Revolutionary Solidarity Movement (1980). Soviet armour penetrated along the main roads into Budapest in 1956 without effective opposition, and seized the Danube bridges (each of which could be quite easily covered by a single tank), thus splitting the insurgents into isolated 'strongholds'. These had similar characteristics to the Paris 'internal fortresses' of 1871. They were usually barricaded monumental public buildings (such as the Kilian, or Budaorsi Road barracks; the Corvin cinema; or the two main railway stations) especially where physical features (such as the Gellert and Varhegy Hills, or Csepel island) hindered the approach and field of fire of armoured vehicles (see Lomax 1976; Irving 1981).

Most examples of open, urban insurrection against regular forces resulted, like Budapest in 1956, in gallant failures to fuel revolutionary mythology rather than the successful overthrow of governments. In Vienna in 1934 the working-class residential blocks of Heiligenstadt proved defensible, at least for a time, against the armoured cars and light artillery of the Austrian army.

Barcelona provides a number of examples of what were effectively civil wars between various factions, culminating in the 1937 struggle between communists and anarchists, with both sides being effectively insurgents in terms of their organization, equipment and tactics. The successive Warsaw uprisings, that born of despair in the Jewish ghetto in 1943, and that of the much better prepared and reasonably armed 30,000 strong 'home army' in 1944, had little chance against the Wehrmacht (Garlinski 1985). The support of regular forces expected by the Poles in 1944 did not materialize, although risings elsewhere in Poland as part of 'operation Tempest' (such as in Lwow, Lublin, Wilno and Bialystok) did co-operate with regular Red Army units. Similarly, in the Slovak uprising of the same year (concentrated initially in the towns of Banska, Bystrica and Zvolen), co-ordination of guerrillas and regulars was successful. In more recent years it is possible to see the North Vietnamese/Vietcong Tet offensive of 1968, as the attempt to extend a mobile rural war into the cities, especially Saigon and Hue. But not only were the forces of both sides principally regular, even if the tactics (especially in the steets and walls of the old city of Hue) were not, the offensive can hardly be counted as a military success, regardless of its political implications (Karnow 1984).

Indeed, the modern archetype of open insurgency is not that of the Paris Commune, where citizens take to the streets in response to the ideological leaadership of the revolutionary cadres, defeating (after a brief struggle on the barricades) the alienated forces of a demoralized government as the culmination of the evolution of insurgency through the various stages outlined earlier. A more realistic model of 'success' is provided by Beirut, which, together with its side-shows in Tyre, Saida and Damour, is the only major city in the modern world where open insurgency has been maintained over a substantial period of time. (It is probable that the United Nations' peace-keeping force has saved Nicosia from a similar fate since 1974.)

The battlefield is composed largely of modern concrete and masonry buildings which have proved remarkably effective in providing firing positions for small arms, and none of the contending insurgent armies possess quantities of artillery, rockets, mortars, or ground–air support, to which alone they are vulnerable. The disintegration of the regular Lebanese army early in the conflict and the lack of decisive intervention by other regular armies (whether Syrian, Israeli or multi-national peace-keeping forces) guaranteed no effective regular opposition. The military balance between the various insurgent factions supported by their respective social groups, awards victory to none, allowing an almost permanent state of chronic disorder to be maintained.

In its first phases, from 1975, the conflict was between Maronite and Muslim militias, which stabilized along the 'green line' splitting Muslim west from Christian east Beirut. Conflict occurred (Figure 4.2) either in the enclaves located on the 'wrong side' of the line, which were 'cleared out', or in the

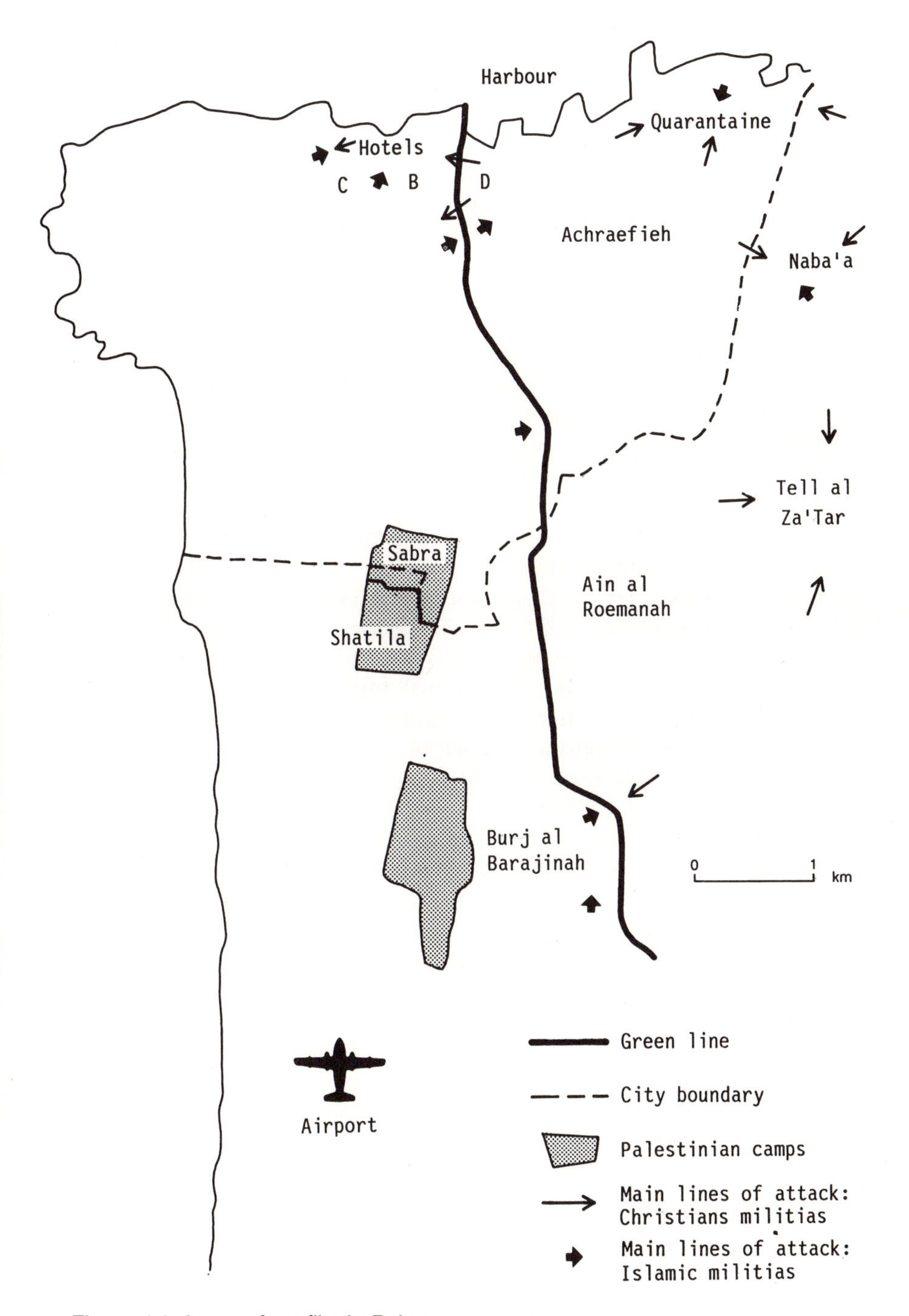

Figure 4.2 Areas of conflict in Beirut

no man's land of the central commercial area, especially the 'battle for the docks', and the 'battle of the hotels' (Gordon 1980). The intervention of the Syrian army in 1976, and the 1982–3 Israeli invasion complicated the situation. Both of these forces were dependent upon armour, and the latter also made extensive use of air-ground support. The Israelis made no serious attempt to occupy Beirut, let alone pacify it, being content to shell it and other military positions in support of their local client militias. The Syrians, on the other hand, in pursuit of a peace-keeping role, used armour to dominate the main roads into Beirut (between the suburban residential districts and the southern airport) but were largely incapable of penetrating into the residential districts themselves. In this phase of the conflict, the southern and eastern suburbs (the 'misery belt', which had been rapidly developed in the 1960s and 1970s to house rural–urban migrants, and become swollen with refugees both from Palestine and from enclaves elsewhere in the city) became the principal battlefields (Ellis 1983). The existing social segregation – into not merely Muslim or Christian but into particular religious-political adherence (whether Amal or Hezbollah, among Shiites, Falangist or Chamounist among Maronites, or various Palestinian factions) – was intensified, and the morphology of close-packed streets and alleys favoured the defensive capabilities of small groups of local militia with small arms (see Mercillan 1982). The combination of a balance between the various internal factions whose offensive capabilities are limited, together with a reluctance of the regular external forces to become too deeply embroiled, has provided a sort of permanent insurgency – with no sign, as this is being written, of resolution after 14 years of conflict.

CONCLUSIONS

The long hisory of urban disorder makes it plain that cities have always been the focus of violence, and that the maintenance of a minimum of public order, without which the other functions of the city cannot begin to be effective, has always been a major preoccupation of urban governments. It is therefore all the more curious that the many studies of urban origins, morphogenesis, and of the economic, administrative and social functioning of cities, should pay such scant regard to the order that is a prerequisite for the operation of these functions, and the perennial threat to that essentially fragile order offered by insurgency.

Two countervailing tendencies, both of which are central to late twentieth-century life-styles, come together in the modern city. On the one hand, the growing complexity of urban life, and the increasing application of inter-dependent technologies needed to make that life possible, increases the vulnerability of the city and its citizens to disruption, while on the other, the same advances in technology and organization increase the effectiveness of the city's capacity to protect itself against such threats. The result is an equilibrium, as it generally always was in the city throughout history, where disorder is contained, although at an increasing cost commensurate with the increasing size, importance and complexity of cities.

There is no definitive urban geography of insurgency, in the sense that the incidence of insurrection can be causally related to particular social and economic characteristics of cities, nor can the chances of success be predicted from socio-economic statistics. Still less is a clear spatial ecology of insurgent areas within cities recognizable. However, there is an important, clear and intimate triangular relationship between operational tactics, the characteristics of weapons and weapon carriers, and many physical attributes of the urban scene. The significance of the spatial patterns of streets and buildings, and the materials of which they are constructed, has been demonstrated in many of the cases described above in relation to the tactics and equipment of the participants. It is equally clear that the incidence and course of insurgency is also influenced by such a wide variety of other organizational, ideological, personal, and mere chance variables (operating both inside and outside the city) that morphological and terrain factors have rarely been decisive in the success of either insurgency or counter-insurgency.

If success is measured in terms of the permanent overthrow of the established order and its intentional replacement by another, then insurgent operations have few successes to their credit. These few, which are often used to inspire others, are, more often than not, cases where external factors intervened. The future of insurgency is also not promising, according to one influential commentator for whom 'the age of the guerrilla is drawing to a close' (Laqueur 1977b), and the revolutionary is retreating into urban terror and crime. If, however, success is taken to mean influencing urban behaviour, urban design and, ultimately, the quality of urban life, then insurgency in all its forms – from street crime, demonstration and riot to terrorism and guerrilla war – has been, and is likely to remain, a chronic condition, and thus a formative influence upon cities.

Chapter 5

The city as battle terrain

A TERRAIN TO BE AVOIDED

The use of the city, as a battlefield has been implicit in the preceding chapters. This urban function, providing places where battles are fought, has such an obvious and fundamental importance to cities, that it should now be made explicit and be considered from the point of view of those attacking or defending them.

A major difficulty in attempting this is the deeply entrenched military opinion, that goes back many centuries, that cities are places where battles should not be fought. Consequently, when it occurred in urban areas, conflict tended to be regarded as an unfortunate aberration to be avoided in future, rather than an example to be analysed so that lessons for the future could be drawn. The literature on military practice, although generally full of advice on the choice of battle sites, has therefore tended (with very few exceptions) to avoid this class of terrain altogether.

Throughout most of recorded military history, warfare could be divided into two clear arenas of action: battles and sieges. The first was undertaken by 'field' armies (the choice of terminology is itself revealing) which manoeuvred against each other until engaging in selected open terrain, chosen by one or both parties. The resulting battle may then have decided the fate of countries, empires, dynasties, and the cities over which they ruled. The second was a static confrontation in which fortified cities were besieged and blockaded. The outcome, whether the surrender of the town or the lifting of the siege, was generally determined without fighting within the city itself.

The urban siege

Fighting in the city itself was not in the interests of the civilian inhabitants of such a battlefield (whose lives and property would inevitably suffer even if they were not the principal objective of the attack). Nor was it of interest to the military authorities on either side, who regarded fighting in the city as dangerous, unpredictable in its outcome and duration, and unsuited to the organization, training and weaponry of most armies, as well as wasteful of resources. A devastated city contained few rewards of victory.

Such a mutual aversion to fighting in the city (for sound practical reasons, shared by attackers and defenders alike) was elevated to a convention of military society. This convention was incorporated into military codes of practice, which transcended dynastic or national allegiance as part of a semi-mystical, professional honour system, linked to a warrior class (Brand 1989). Such codes were adhered to most strongly when warfare was dominated by such a professional class with such a professional ethic. In particular, this formal morality was the philosophical counterpart in the 1500–1800 period to the 'military science' approach to urban defence and attack. Although weakened by the rise of nationalism – which undermined the international 'brotherhood of arms', and the replacement of professional by popular war – such codes reappeared in formal international agreements on the treatment of cities, such as The Hague conventions at the end of the nineteenth century.

Cities which chose not to defend themselves at all could be granted the privileges of 'open cities', which at best might include security of property, life and liberty for civilians, and withdrawal, with the 'honours of war', for defending military personnel. Sieges were conducted according to 'rules' which required surrender after so many days, or by mutual agreement if relief was not forthcoming within a set period, or after an agreed portion of the fortifications had been seized by assault or breached by artillery. It was generally enough to demonstrate that the city *could* be taken without the cost to both sides of actually taking it. The defence of the indefensible was not permitted. 'The defence of a fortress which was duly summoned, and then stormed had no claim to mercy, the more so if the fortress was patently indefensible' (Cromwell (1649) as quoted in Buchan 1934: 352). Further resistance within the city itself, after the fortifications had been breached or taken, thus causing the attacker to engage in urban warfare, not only encouraged attacking forces to claim pillage as compensation for the costs of urban fighting, but was tantamount to abandoning all the conventional rights of quarter.

Deliberate breaches of these conventions fill the history books and are the stuff of which popular mythology is made. Their longevity in folk memory is a result of their being exceptions whose very rarity shocked contemporaries and guaranteed a degree of immortality for such events. Typical examples of the memorable, and thus unusual, quality of the sacking of cities and the killing of their civilian inhabitants after siege are as follows: the destruction of Carthage in 146 BC or Jerusalen in AD 70; the pillaging of Naarden and Haarlem by Alva during the Dutch revolt in 1572; the sacking of Magdeburg by the imperial army in 1631; Cromwell's 1649–50 campaign in Ireland and the massacres of Drogheda and Wexford. All these examples, and the many more before and since, contain the same elements of a desire to deliver a memorable shock on the part of the perpetrators so that the setting aside of the conventions of urban warfare would lead to a quicker, and permanent, end to the campaign, through the immediate surrender of subsequent besieged cities.

Urban destruction as part of a campaign strategy has a long lineage – from the Israelites' massacres of the inhabitants of the Philistine and Canaanite cities, to Hiroshima (see Chapter 6). In addition, the 'professional' conventions depend upon the maintenance of a clear distinction between soldiers and civilians, and the adherence of both to a mutually accepted code of practice. Civil war or popular insurgency removes the military/civilian distinction, while wide or strongly held differences of religion, ideology or culture destroy the link between opposing professionals, and generally render such codes untenable.

The urban battle

The choice of urban sites as places to engage and destroy enemy forces, as opposed to sieges where the objective is the city itself, releases the participants from the various codes and conventions described above. Despite the memorable exceptions (which will be referred to below) it is the relative paucity of examples of conventional forces engaged on urban battlefields that remains notable. In addition to the various professional and technical arguments to explain this distaste of the military for urban battle sites, and the inevitable complications introduced by the presence of civilians – who, even if not especially protected, nevertheless quite simply get in the way – an additional obvious but important explanation lies in the degree of urbanization. Towns were both scarcer and smaller until the urban revolution of the nineteenth and twentieth centuries, so that the possibility of using them as battlefields was simply rarer.

A high proportion of the examples that can be found are conflicts in which one of the participants has the characteristics of an irregular force of one sort or another.

The armed revolt of the Flemish towns at the end of the thirteenth century was exceptional, in the opposition of an urban commercial and artisan population to a feudal overlord, leading to the expulsion of feudal heavy cavalry from Bruges in 1297 – in street fighting with local militia. It is true that, in the Flanders of Jan van Artevelde, the size, number and density of urban settlements, the commercial development that supported them and the social changes this had wrought upon the inhabitants, were all themselves exceptional in late medieval Europe – and even here the decisive (1302) Battle of the Spurs was fought outside not inside the walls of Kortrijk.

The two successive sieges of Paris in 1870–1 are dramatic and contrasting illustrations. In the first and conventional siege between regular armies, the city was surrenderd as a consequence of battles outside the walls, without a

German solider entering the city, and no reprisals were taken. In the second siege, fought between a professional and an insurgent force, the critical fighting occurred within the city (see Chapter 4), and the conventions of civilian rights were subsequently widely ignored by both sides.

The distinction between urban warfare and police actions by regular forces in urban areas can be blurred, and many such actions since the Second World War are more appropriately described in Chapter 4. However, some examples – such as the 1945 action between British Indian Army units and Indonesian Republican forces in Surabaya – clearly approached the warfare end of the spectrum in scale and operational organization.

Cases where both sides were regular professional armies are even rarer. Many of the pre-nineteenth-century examples of armies clashing within towns are both unintentional and little more than skirmishes if one looks at the numbers involved. Winchester (1141) was the case of a force besieging the castle being itself besieged, and thus inadvertently engaging in urban warfare, while Lewes (1264) was a case where cavalry overshot its original open target and continued a pursuit into the streets of the town. Many of the battles of both the Wars of the Roses and the English Civil War that carry the names of towns (whether Tewkesbury and Northampton in the former, or Newbury, Worcester and Dunbar in the latter) in fact occurred in open country outside these towns. Only at St Albans in 1455 and at Salisbury in 1644 was the urban morphology an intentional aspect of the conflict, both being struggles around a created defence line – composed of street barricades in the former case and the cathedral close in the latter. One of the last battles fought in Britain (namely, Preston in 1715) is one of the few examples from British history where both sides were composed of regular soldiers. Here Jacobite forces converted the town into a fortified position by street barricades covered by musketry loopholed buildings, that were eventually taken by costly frontal assault (Green 1973).

Even in the period since the Napoleonic rediscovery of mobile warfare and the end of the period of dominance of campaign strategy by the fortified city, there are few cases to be found. For example, the many Napoleonic campaigns contain no sizeable urban battles, despite Bonaparte's rise to fame as an exponent of the use of artillery in the suppression of urban revolt in 1794. Moscow was neither defended by the Russians in 1812 nor subject to the scorched-earth policy practised in the countryside, but simply abandoned. Only in Spain, where the conflict had many of the elements of a civil war, was there street fighting – such as in Madrid, Badajoz and other towns in 1808.

Even twentieth-century warfare, fought over an increasingly urbanized world, provides remarkably few examples of the deliberate choice of an urban battlefield, although these few were often memorable. Stalingrad occupies a unique position in the mythology of the Second World War, although it is at least arguable (Erikson 1975) that it was not essential for the

German offensive of August 1942 to occupy the city, nor for the Russians to continue to defend their west bank of the Volga bridgeheads in it. The fate of the German Sixth Army was sealed by the Russian offensives of October/November outside the city, and the subsequent Russian attacks into the city (December 1942 to January 1943) served little military purpose. Clearly, both sides had ascribed symbolic values to the place itself. This element can also be detected in other major urban battles of the war, such as Rostow on the Don (1942), the Normandy towns of Caen, Ouistreham and Bayeux (1944), Budapest (1945) and Berlin (1945). Equally, these were all occasions where defensible terrain had been selected by the defenders. Among the cases of attacking forces choosing an urban environment are the German air-landed offensive in the Western Netherlands in 1940 (see de Jong 1970) and the Allied Arnhem/Nijmegen airborne assault in 1944. In both, the objective was to seize bridges that happened to exist in the cities rather than the cities themselves.

However, the length and importance of such a list is less notable than the many absentees from it. In particular, conventional military wisdom by the 1940s (inspired in part by the defence of Madrid and other cities in the Spanish Civil War) had begun to regard urban environments as providing good defensive terrain, especially for infantry faced with numerically superior, more mobile or armoured forces. However, although finding themselves in just these tactical circumstances, these defensive possibilities of the urban environment were deliberately eschewed by the hard-pressed allies: in the France/Belgium campaign in 1940; in the Philippines (1941/2), where Manila was declared an open city; and in the Malay peninsula where the 'fortress' of Singapore was surrendered (1942) before fighting had occurred in the city itself (Turnbull 1977). Even the hard-fought Leningrad campaign (1941–4) was a seige rather than an urban battle, with only a part of the fighting occurring in the outer suburbs and smaller surrounding towns. Similarly, retreating Axis forces made no attempt to fight defensive battles in Tunis, Naples, Rome, Paris or even with much seriousness in Vienna and Prague.

More recently, the North Vietnamese and Khmer Rouge 1975 offensives, which marked the end of the long-fought Indo-China conflict, did not culminate in battles within the cities of Saigon or Phnom Penh respectively, although there were sharp delaying actions fought by the ARVN in some of the provincial towns such as Phuoc Binh, and especially Xuan Loc.

THE CHARACTERISTICS OF THE URBAN ENVIRONMENT AS BATTLEFIELD

The most important fundamental military characteristics of warfare in cities, whether strategic (in modern terminology, MOUT – military operations in urban terrain) or tactical (FIBUA – fighting in built-up areas), stem directly from the nature of the urban environment. These can be summarized for most periods of history under five headings.

Small operational units

The streets and building-blocks of the urban physical morphology fragment urban warfare into conflict between units usually of squad or platoon size, with generally insufficient space for the deployment and manoeuvring of larger units. Complexes of buildings become (or can be relatively easily converted into) defensive positions, and the battle rapidly disintegrates into a series of more or less separate and isolated conflicts around such 'fortresses'. Such positions may be formed from a wide variety of urban structures and materials. In Salisbury (1644) it was a cathedral close enclosed with a masonry wall for musketry positions; in Paris (1871) it was loopholed, large, stone public buildings (such as the Luxembourg Palace and the Pantheon) or brick-built three-to-four-storey apartment blocks linked by barricades of paving stones across narrow streets (Jellinek 1971); while in Stalingrad (1942) it was the massive concrete blocks of the factory district that provided cover from artillery. Even relatively minor features – such as the elevations of the railway sidings at Stalingrad – provided a compartmentalized grid of infantry positions, nicknamed 'the tennis racquet' (Erikson 1975). Ironically, it was a similar combination of restricted river crossings and factory architecture on Csepel Island that proved to be the most difficult terrain for Red Army armour during the suppression of the Hungarian uprising in 1956 (Lomax 1976).

A consequence of this fragmentation is a loss of control of the battle by commanders and resulting devolution of responsibility to small group leaders and individuals, which accounts in part for the distrust of a professional officer corps in committing forces to a battle over whose progress they have minimal control. A clear, and for the attacking forces disastrous, example of this loss of command control occurred in the American attack on Quebec City in 1775, which degenerated into a confused series of night-time skirmishes around street barricades which resulted not only in the failure of the American commander to deploy his superior forces and take the city but his death and the collapse of the offensive (Stanley 1973). Similarly, Warsaw's 'Home Army' uprising of 1944 (Garlinski 1985) was hampered throughout by a lack of communication between insurgent units, which, once committed to an area of the city, were largely outside command control, whereas the Wehrmacht maintained better communication and thus freedom of deployment of (not greatly superior) numbers.

Conversely, such conflict will emphasize individual motivation, reliability and initiative, and the quality of junior leadership, rather than the wider regulatory disciplines and *esprit de corps* upon which most professional armies are based. These conditions, as argued in Chatper 4, favour the individually motivated guerrilla over the technical and organizational skills of the professional; and the personal skills and self-reliance of such 'regular irregulars' as paratroops, commandos, 'green berets' and the like, rather than units trained for more conventional warfare.

The most fiercely-fought urban battles were generally conducted by units with these characteristics. Stalingrad pitted a Wehrmacht in which particular attention had been paid to the quality of small-unit leadership, against a Red Army which stressed the importance of individual motivation. Similarly, in the urban examples from the series of Arab–Israeli wars since 1947, the Israeli Palmach and such semi-official 'armies' as the Irgun were individually motivated and trained to operate as small units. They had little difficulty 'clearing' Haifa, Jaffa and Acre in April 1948, mostly by a process of seizing strategically important buildings and elevated positions commanding access to and surrounding the densely built-up Arab cities (thus provoking Arab civilian panic and surrender) rather than by house-to-house penetration. However, they were unable to hold East Jerusalem against a Jordanian 'Arab Legion' – in which professional training as well as the social mores of the Bedouin from which it was recruited, emphasized the small unit – and a position of stalemate was reached. The June 1967 resumption of hostilities led to a fiercely-fought infantry battle around the medieval walls and gates of the old city, largely between Israeli paratroops and Arab Legion infantry, with the ultimate victory of the former (Herzog 1982).

Close-range weaponry

A fundamental feature of the urban environment is that buildings foreclose visibility. The most important consequence of this simple condition is that weapon ranges are necessarily short. Currently, NATO forces work on the assumption of a '3000m battlefield' (Kamps 1980) which grossly overestimates most urban conditions for which US infantry training manuals (quoted in O'Sullivan and Miller 1983) estimate actual ranges, even in suburban conditions, to be between 125 and 250 metres. The Soviets largely concur, estimating that 40 per cent of all engagements will be fought at ranges of less than 500 metres, and almost two-thirds at less than 1 kilometre.

These short ranges have a number of consequences. They are too short for the safe operation of much heavy weaponry, and confine the bulk of urban fighting to hand-held, or hand-thrown infantry missiles. In addition, most artillery, armour, and supporting air or sea ground-support is generally unable to sight targets, which, even if visible, are difficult to identify in densely-built-up areas. Indeed, it can even be counter-productive, in that the attempt to hit targets in the city creates 'collateral damage' over a wide area surrounding the target. This, in turn, is likely to reduce accessibility, visibility, and recognition within the city by blocking roads, as well as creating infantry positions among the rubble, all of which favours defence. The co-ordination of arms, a central aim of advanced military operations, is thus rendered ineffective.

These difficulties of visibility and recognition, which are exacerbated by damage to buildings and streets, frequently result quite simply in units becoming lost in a maze of unfamiliar terrain. Consequently, however advanced the communications systems and the supporting artillery and air power, small units become dependent upon their own resources and fight individual battles without secure flanks, rear or reference to the wider battle context (a situation vividly illustrated in the film *Full Metal Jacket*, directed by Stanley Kubrick, which follows the trials of an infantry squad in the unfamiliar streets of Hue, Vietnam, in 1968).

This situation could also be illustrated by the 1967 battle for Jerusalem, when Jordanian artillery sited on the high ground to the east of the city maintained a constant but military ineffective shelling of the western Jewish suburbs, while Israeli air-power was unable to support their ground forces in the old city, although successfully interdicting Jordanian reinforcements in the open country to the east of the city.

The same part of the world provided a number of examples in the 1980s of the difficulties faced by armour-dependent armies when opposed by lightly armed irregulars in an urban environment. The rapid mechanized advance of the Israeli army through Lebanon in 1982 met its first effective opposition from units of the PLO in the southern suburbs of Beirut. The subsequent retreat of the Palestinians was as much a result of shelling from surrounding positions and pressures from the indigenous inhabitants as effective infantry penetration of the city itself. Three years later in the same districts, a Syrian army (also based largely upon armour) was unable, or unwilling, to impose its presence upon the Amal and Hezbollah militias ensconced in the densely-built-up residential blocks in the south of the city, whose architecture provided restricted fields of fire, and numerous potential close-range concealment and ambush positions (Mercillon 1982).

The presence of civilian lives and property

A fundamental characteristic of the city as battlefield is the presence of civilians unavoidably and inextricably caught up in the conflict. This may impose restrictions on movement, fields of fire, targeting, weapon choice and many other military options. Attitudes towards the city's civilian inhabitants, and thus the severity of the constraints upon the military, depend upon political or humanitarian considerations. In this respect, during the allied advance through western Europe in 1944–5, the liberation of cities inhabited by 'friendly civilians' (in France and the Low Countries) imposed more limitations upon the occupying armies than did the cities of Germany. For example, in the Battle of Groningen (described below) the necessity for infantry street fighting rather than the preferred military artillery option was dictated by this consideration. (Ommen Kloeke 1947). Even when a diversion of military services and manpower is not required for the care of civilians, they will nevertheless impose some constraints upon military action in protection of their lives and property, or just be in the way. This coincidental presence of civilians and accidental involvement in the conflict is exacerbated by some of the other characteristics of the city already mentioned. The short weapon ranges involved – and, thus, fast reaction times – makes it difficult to identify targets correctly and therefore increases civilian casualties. Similarly, regular forces will be aware that the urban terrain is ideally suited to the concealment, ambush and withdrawal techniques of partisans, and a predisposition to regard civilians with suspicion is likely to exist. Allied armies in the 1940 western European campaign had an exaggerated fear of 'fifth columnists' (de Jong 1970), and in 1944–5 were acutely aware of the possible existence of *francs-tireurs*, both of which beliefs were encouraged by enemy propaganda, even in 'friendly' liberated cities (Huizinga 1980). This predisposition towards suspicion will be compounded in units anxious for the security of their flanks and rear and uncertain of their location or the position of other friendly forces.

Thus the presence of civilians on the battlefield at best imposes constraints on the conduct of that battle and at worst leads to a breakdown in the military/civilian distinction, which increases both the hardships suffered by the inhabitants and the difficulties faced by the military.

Defensive bias

Although the circumstances described above apply to both attacking and defending forces, there is reason to argue for an inbuilt bias in favour of the defence. A Soviet military standard, developed from experience in the Second World War, is reported to regard a 3 : 1 attack/defence manpower ratio as a minimum requirement in rural areas, but 10 : 1 in towns (O'Sullivan and Miller 1983).

This stems in part from the characteristics of the battlefield environment itself – with the close density of buildings, and the rubble that they will provide in the course of the battle, providing opportunities for ambush, mining, booby-trapping and the like, all of which are essentially defensive. Such forms offer easily adapted and camouflaged positions (especially for defending infantry), while hindering the visibility and mobility of advancing armoured vehicles – which must approach within range of anti-tank projectiles. Not only do cities favour infantry, which is more likely to be defensive, over armour, which is more likely to be spearheading an attack, but, more generally, they present a circumstance in which troops that are inferior in equipment, training or morale can be pitted against superior forces on more even terms. This was aptly illustrated in the 5-day campaign in the western Netherlands in 1940, when a German battalion of air-landed troops seized the south side of the Willemsbrug over the Maas in Rotterdam on the first day of the invasion and could not be dislodged. Equally, even when reinforced by a Panzer division, they were unable to advance across the Maas to the north side (de Jong 1970).

All defenders, whether in urban areas or not, can be assumed to have some advantage of time to prepare positions, to have a better knowledge of the battle site than attacking forces, as well as a better chance of useful co-operation with local civilians. However, the sheer complexity of the urban network of roads, alleys, passages, subways, sewers, and telecommunication links should convey important advantages of accessibility and flexibility on the forces with a better knowledge of such systems. Defenders should, therefore in most circumstances, have a greater ease of interior communications – thus facilitating reinforcement, tactical withdrawals, infiltration, counter-attacks and breakout offensives.

Absorption of manpower

This last characteristic is largely a resultant of the operation of those already mentioned but remains the most important military consideration governing the choice of cities as battlefields. Quite simply, the amount of resources (especially human resources) needed to conduct an urban battle is extremely high in relation to the area of the battlefield. The urban environment creates a highly physically structured but fragmented series of compartmentalized battlefields that can absorb large quantities of personnel – which, once committed, will be difficult to extricate, regroup or reinforce. Whether casualites will be heavier (as a percentage of the committed manpower, rather than in terms of space won or lost) is arguable, but the tying down of large quantities of troops for long periods and, thus, the delaying of other operations for limited spatial objectives, is not. Similarly, the type of battle dictated by the urban environment imposes particularly severe strains on those subjected to it. The continuous high level of alertness demanded by close actions, the physical discomfort, and the insecurity of isolated small unit operations without fixed lines, secure flanks, or protected rear all contribute to the rapid onset of battle fatigue within hours rather than days.

Keegan's (1976) study of the reactions of men in battle has raised a related point, in order to explain the remarkable ferocity which many have reported to occur in close-combat battle situations. The critical distances observed by animal behaviourists are, according to Keegan, related to the human aggression which is triggered by the close physical presence of the enemy and the personal threat that this implies. If, as he claims, such a reaction can be observed in the closely confined small action in the Hougemont Chateau at Waterloo, then it is likely to be even more applicable when such physical conditions are extended in space and time, as in an urban battlefield. The

resulting ferocity will be reflected in both faster exhaustion and a higher level of casualties, due in part to a reluctance to accept individual surrender. The necessity for the rapid rotation of units (where command control exists) or the addition of reinforcements (where it does not) contributes to the high levels of manpower needed.

The recognition that the urban battlefield is distinctive, and thus requires specialized military skills, may be reflected in specialized training and the development of assault groups. However, such a professionalization of the urban battlefield is itself a cost, not least in unit flexibility, as well as being extremely difficult to achieve because of the special problems of peace-time training in urban areas (see Chapter 3).

URBAN MILITARY OPERATIONS

The effects of these general characteristics of the urban environment as a battle terrain upon military operations in urban areas can now be considered in more detail, drawing upon the very limited range of historical experience available and also upon the future scenarios which have shaped training for such battles since the Second World War.

The strategic scale

Warsaw Pact 'urban doctrine' is among the most clearly articulated set of guidelines for urban operations developed since 1945. A first principle is to avoid major conurbations but, having determined that a defended city is to be taken (rather than masked and bypassed) then the series of actions

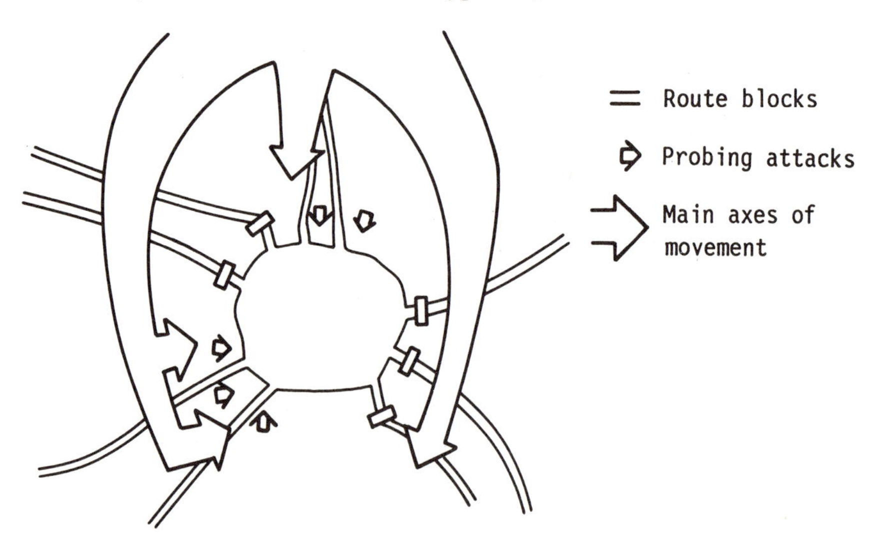

Figure 5.1 Warsaw Pact idealized urban attack

summarized in Figure 5.1 is initiated (Lewis 1982). Although intended for a contemporary scenario, these would not have been unfamiliar in any period since the introduction of mechanization, and have most recently been restated in an only slightly modified form by the Israeli defence forces (Mercillon 1982) as a result of their Lebanon experience. The advanced, fast-moving echelons attempt to overrun the city before preparations for its defence can be initiated. If this rather optimistic possibility fails, then the advanced units bypass the built-up area, seizing (with air-landed support if necessary) critical bridges and road junctions, thus blocking potential lines of reinforcement, resupply or withdrawal, and wait for second-echelon reinforcements. Probing reconnaissance attacks in relatively small units are then mounted to determine the direction and nature of the main assault, which is then undertaken by specialized, motorized infantry units – supported where necessary by tanks held in reserve. The objective is to seize transport, communication and command centres so as to fragment the defence into isolated pockets. Reconnaissance assumes a critical role in mitigating the problems of loss of command control once the main assault is under way and unfamilarity with the local terrain. The synchronization of this external assault with an internal insurgency (aimed at reducing the defenders' mobility, control over the battle, and morale) is regarded as an ideal situation more likely to occur in friendly than hostile territory – as reported by Rustin (1980) for current Yugoslav urban defensive practice.

Given the assumed stategic objectives in the European central front during a potential major conventional conflict, Warsaw Pact guidelines for the defence of cities were much more rudimentary, amounting to little more than a resolution to fight outside the city and to prevent the initial envelopment. The emphasis in NATO doctrine has been the mirror image of the Soviets, based on very similar preconceptions. The cities, and especially the densely-built-up central areas of the large West German conurbations (such as Hanover and Braunschweig), are to be used as defendable strong-points. Thus the assumed Soviet numerical superiority in armour is to be neutralized by the deployment of infantry, including local territorial units, in a series of delaying actions, which blunt and absorb the Soviet thrusts for long enough to allow either inter-continental reinforcement ('Operation Reforger' and the like) or political resolutions to occur.

Such scenarios dominated strategic thinking in Europe between 1950 and 1990, and accorded a special importance to towns. Changes in European security since 1990 will lead to new strategies and scenarios but, in an urbanized continent, such strategies are unlikely to diminish in importance.

The tactical scale

The examination of the actual effects of the character of the urban environment upon tactical operations requires a degree of detail about the use of urban space and the structure and materials of the urban battlefield that is often missing in accounts of the fighting. Given the essentially confused, small-scale nature of urban conflict, it is also difficult to assess.

One closely-worked example, so constrained in time and space as to be understandable – and for which such details exist (Ommen Kloeke 1945; van Welsenes 1976; Huizinga 1980; Tammeling 1980), or can be deduced – can serve as an illustration of the effects of the urban environmental characteristics (argued above) and the tactical reactions of defenders and attackers to them. This illustration should also be applicable at other spatial scales and at other times, at least since the introduction of firearms.

The Battle of Groningen (The Netherlands)

Context

This battle occurred over the four days 13–16 April 1945, at reinforced infantry-division strength without the appreciable intervention of other arms, in a suburban and inner-urban environment. The wider, strategic context was the decision of Allied Twenty-first Army Group to advance from the Rhine to the North Sea Coast through the north-eastern Netherlands, thus isolating substantial German occupying forces in the western Netherlands. The principal German obstacle, and the main objective of 'Operation Plunder', was the 'fortified' city of Groningen, which was the focus of the road, rail, and water transport systems, and was difficult to bypass, not least because the age-old Dutch defensive measure of strategic inundation channelled the advance along a limited and predictable number of corridors.

The city region had a population of around 200,000 in a relatively compact built-up area whose inner city was composed of a high density of 2 to 5-storey, principally brick, building on a dominantly Late Medieval street pattern. The important interruptions to this pattern of short narrow streets were the open spaces of the three principal markets, including: the centrally located Grote Markt; the long stretches of 'boulevards' along the line of the seventeenth-century fortifications, especially on the southern border of the inner city; and the few broad streets laid out along reclaimed waterways. The outer areas consisted largely of ribbon development along the southern access roads (either detached houses and gardens, or, closer to the city, 4 to 5-storey, inter-war, brick apartment blocks directly fronting the street). Its 'natural' defences – canalized waterways and narrow, 'dry' access routes separated by 'wet', agricultural land criss-crossed by drainage ditches – had been strengthened after June 1944 by wired weapon pits along the 'inside' canal banks, masonry bunkers covering the main bridges, tank barriers across the two roads to the south, and even an attempt to dig a (partially above ground) 'trench system' in the low lying open land between the main roads to the south of the city (van Welsenses 1976).

Opposing forces

The attacking forces were drawn from the Second Canadian infantry division. The Groningen battle honour was awarded to no fewer than thirteen Canadian battalions, of which ten were infantry battalions, supported by squadrons of Sherman tanks and units from two armoured-car reconnaissance battalions. However, these infantry units were generally rotated during the battle on a daily basis, as a result of the 'battle fatigue' factor, so that individual units rarely engaged in more than 12 hours of street fighting in the inner city. Thus, total attacking forces numbered around 6,000–7,000 high quality infantry (not all of which were in action simultaneously, but some of which were in armoured troop carriers and equipped with flame-throwers) supported by tanks armed with 75-millimetre guns.

The composition of the defending German forces is less certain but is believed to have numbered around 6,000. This seemingly very favourable defence/attack ratio must be qualified by the very mixed quality of the defenders. A high proportion of the total could not be counted as combatants at all, the rest included German police units, railway personnel, and local SS personnel as well as Wehrmacht garrison troops. About twenty multi-purpose, 2-centimetre 'flak' guns and infantry anti-tank 'panzerfausten' were available.

Course

The approach phase, 13–14 April 1945 (Figure 5.2) The presence of friendly civilians in an allied city was a major constraint on the choice of operational plan. It ruled out the use of heavy artillery or air bombardment before or during the battle, committed infantry to a house-by-house operation, and rendered the whole eastern flank of the city (where a major hospital complex was located) unassailable, but had the advantage that detailed knowledge of defence preparations and of the layout of the street and building patterns were available. The advance from the south was along both main access roads: the prepared defences along these roads (consisting of concrete anti-tank obstacles) were breached easily, but progress was hindered by defending infantry in the buildings of the suburbs of Haren and Helpman, and in the Stadspark area, threatening the road. In practice, few of the fixed, prepared, defensive positions had much value, apart from channelling the lines of advance, and it was the buildings themselves which proved to be defendable obstacles. In the outer area these included the suburban development directly flanking the road and two massively-constructed public buildings, the main electricity generating station and the sugar beet factory – which were occupied, and thus became fortresses which threatened to interdict the lines of advance.

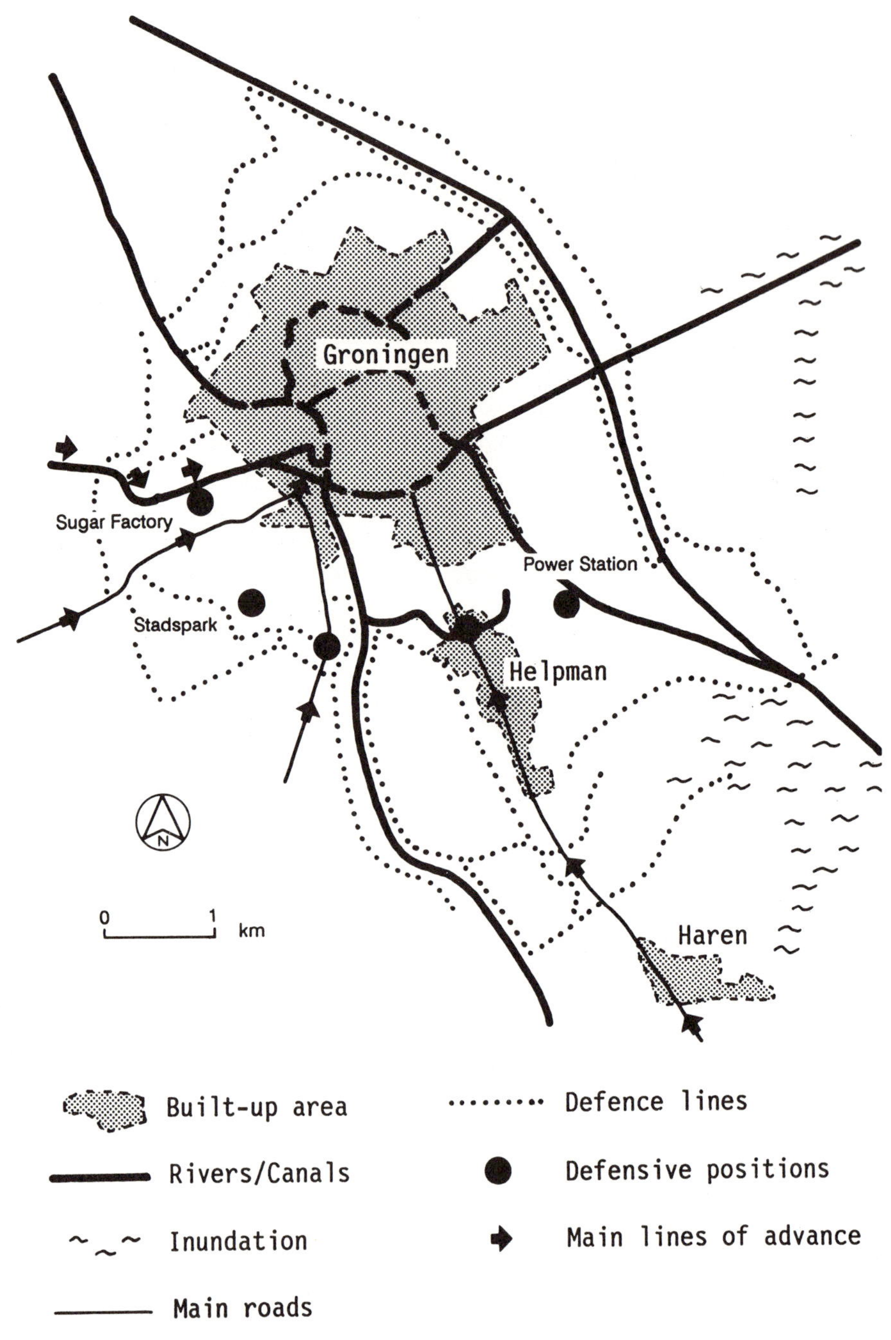

Figure 5.2 Battle of Groningen: the approach

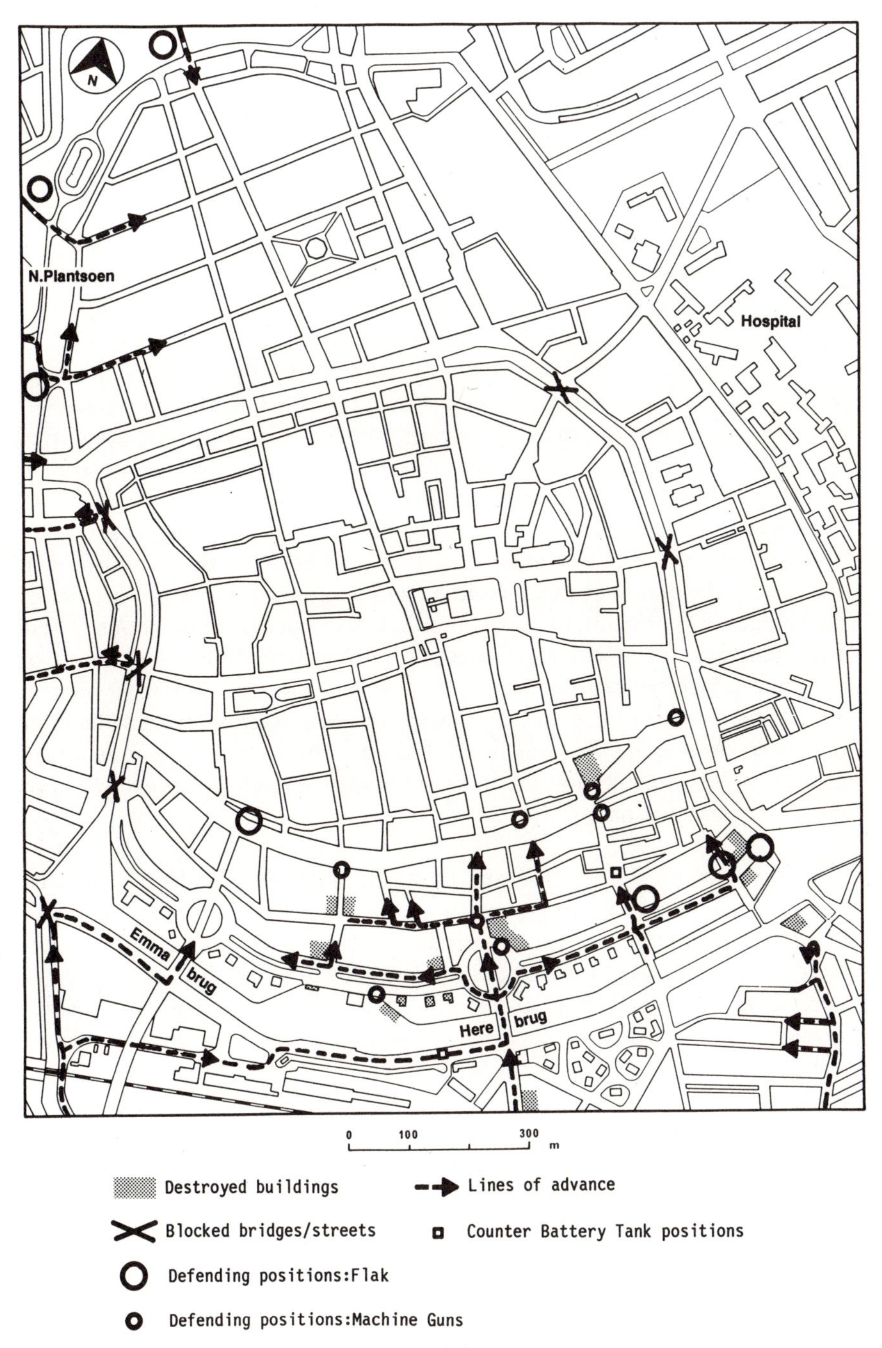

Figure 5.3 Battle of Groningen: the break-in

The break-in phase, 14–15 April 1945 (Figure 5.3) Once the inner city was reached, the battle fell into two phases: the break-in and subsequent clear-out. The extensive water system (including the complete, circular grachten) focused attention on the bridges, most of which had either been destroyed or, more simply, raised. Some, however, remained in place – not least to allow the withdrawal of defending forces from the southern suburbs, and to keep open the option of a retreat of the whole garrison to the remaining German positions in the north-east Netherlands. The break-in occurred at two separate points, which hereafter split the battle into two uncoordinated parts. In the south, the Herebrug, although under fire from strong defending positions in the buildings fronting it, was seized with the aid of supporting armour firing across the canal. In the west, a day later, the railway viaduct was used to penetrate the north-eastern flank of the city and force the fixed positions in the Noorderplantsoen (itself marking the line of the seventeenth-century fortifications), again with a combination of infantry assault assisted by flame-throwers and armour support. A third thrust was successful in crossing into the nineteenth-century district of Schildersbuurt but could not proceed further, since the only intact bridge was too light for tanks.

All of these attacks contained a large element of opportunism and initiative at the platoon and squad level, and demonstrated the importance of the bridges in funnelling the lines of attack on to prepared positions. These were generally for machine guns, supported by 2-centimetre flak units, or anti-tank panzerfausten (whose most effective range was about 80 metres) with suitable fields of fire. Equally, however, such positions were vulnerable to counter-fire (along the same fields) from armour, whose technique was to destroy or set on fire the buildings surrounding such positions and thus cause their evacuation.

The clear-out phase, 15–16 April 1945 (Figure 5.4) The clear-out phase was a fiercely-fought house-by-house 2-day battle along two general lines of advance towards the Grote Markt (namely, from the south and the north-west). The essentially medieval cadastral pattern of a large number of narrow streets, with few straight stretches, resulted in a fragmentation of the advancing forces, and in firing ranges of generally less than 50 metres. These features effectively reversed the procedure of using armour, followed by infantry in favour of small groups of infiltrating infantry, calling up tank support once a street was clear. The main obstacles to this slow but steady advance were the few broad streets and market squares which provided the opportunity for establishing defending fields of fire, across or along them, from prepared machine gun or flak positions. In the south there was the broad, curving Zuiderdiep (a filled-in canal) covered by artillery at each end; the A-Kerkhof/Vismarkt, which surprisingly was not determinedly defended; and, above all, the wide expanse of the Grote Markt itself, which was defended by a series of strong positions along its northern and eastern

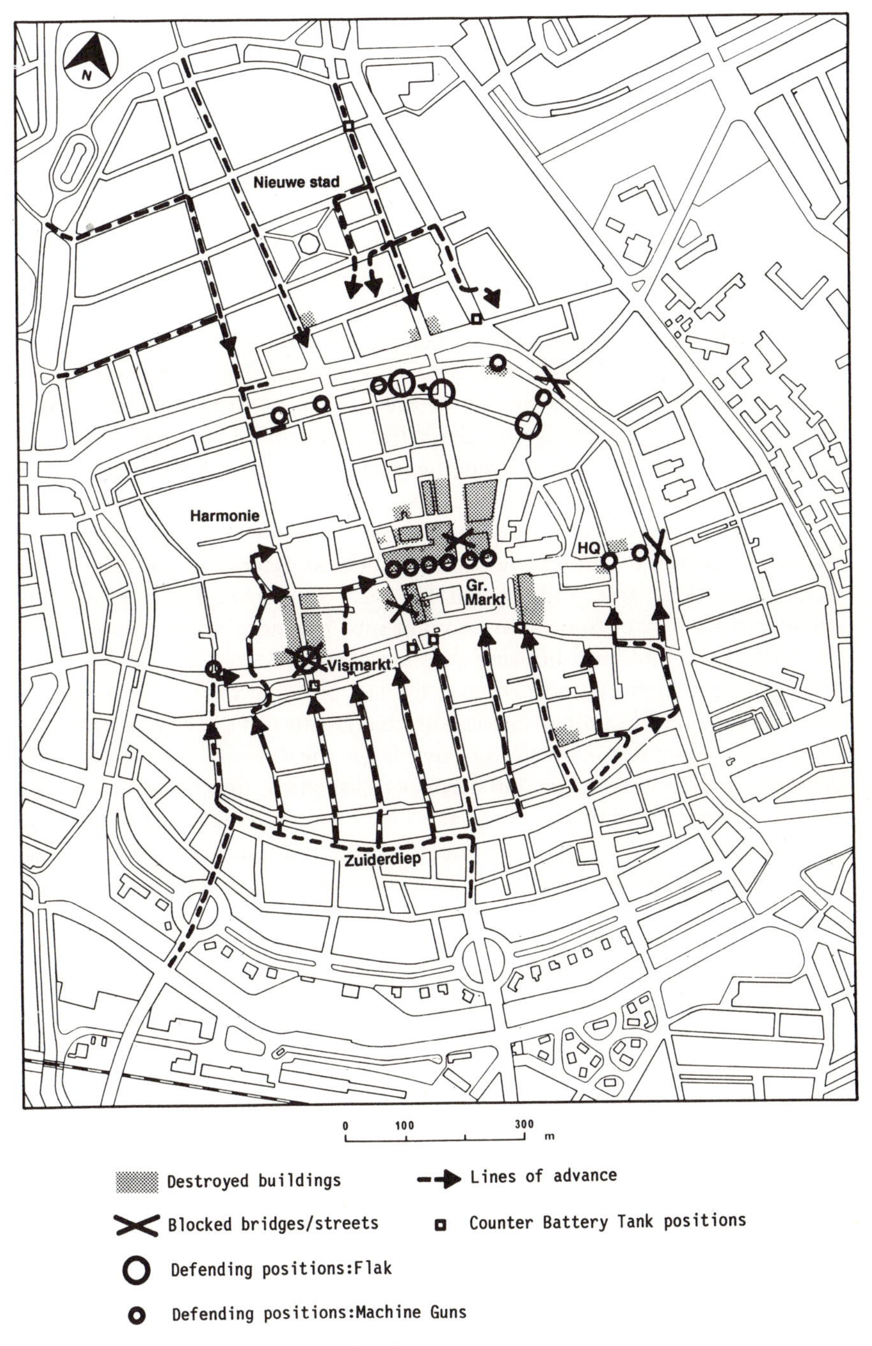

Figure 5.4 Battle of Groningen: the clear-out

edges. Infantry emerging on to the south side of the Grote Markt from the many narrow streets were incapable of proceeding further, and could only infiltrate sideways into the buildings on the southern side and engage in a 2-day fire fight across the square at a range of about 150 metres.

In the northern sector of the inner city, the street plan is subtly different – this being a 'new town' laid out in the middle of the seventeenth century and dominated by a number of straight north-south roads. This provided the opportunity for defending fire up the full length of these roads from positions in the old town on the south side of the ring canal, while, conversely, fire from positions in the 'new town' could not penetrate the streets of the inner city.

The north-east quadrant of the inner city therefore provided a strong and coherent series of defensive positions: covered on the south by the open space of the Grote Markt; the east by the ring canal with no usable bridges; and the north by the ability to direct fire along the only lines of possible advance. This perimeter was strengthened in the course of the battle by the deliberate demolition of buildings to block some of the narrow streets offering possible lines of infiltration on the west side. Resistance was overcome in part by infantry infiltration behind defending positions (often through gardens and walls, aided by local informants) and in part by counter-battery fire from tanks brought into positions where fire could be directed at the buildings housing defending units.

This type of fighting, occurring in the last phase of the battle, resulted in the destruction of buildings in concentrated clusters in the city centre rather than more generally – as would have been the case if heavy artillery or airpower had been engaged. In terms of casualties, figures are vague, but both attackers and defenders appear to have suffered around 300 dead and seriously wounded, which is around 10 per cent of German forces and half that for the Canadians, whose units were rotated in action. More surprisingly, less than 100 civilians were killed, despite the fact that the battlefield was inhabited throughout.

This last phase underlined two interrelated points I made earlier, concerning the value of mobility within the battlefield and information about the course of the battle. On the German side, positions were held until taken or surrendered; forces, once committed, could be neither reinforced nor withdrawn. The advancing Canadians had better mobility, especially noticeable in their ability to summon tank support for infantry at critical points. The Canadians were equipped with staff maps that included street patterns, and information about detailed access possibilities was readily supplied by local inhabitants on a number of occasions. German positions, however (most of which were hurriedly prepared), were not known. An example of the consequences of this ignorance was the erroneous belief that the western side of the German perimeter was protected by substantial forces defending the Harmonie building. The situation on the defending side, however, was far worse – with command control being effectively lost from the moment of break-in. The failure to reinforce the southern bridges, or to defend the open space of the Vismarkt, is explainable by the absence of information. The German commandant when called upon to surrender on the morning of the 16th was unaware of the course of the battle elsewhere in the inner city.

THE DESIGN OF THE FUTURE URBAN BATTLEFIELD

Although cities have not, for the various reasons argued, provided most of the battlefields of most of history, there is a modern consensus that future conflicts will increasingly occur in urban environments. This is a result quite simply of the very rapid and extensive world-wide urbanization that has occurred since 1945, together with the special strategic importance of one of the world's most strongly urbanized regions – namely, central and north-western Europe. During the post-war decades this region was the location of the most important direct superpower confrontation, as well as being an economic and political regional power in its own right. Recent changes in the appreciation of future threats to European security have not diminished its strategic significance. Given this importance and the increasing threats of insurgency, of one sort or another, in cities (as discussed in Chapter 4), it is not therefore surprising that speculation about, and preparation for, future conflicts between coventional forces should also be increasingly centred upon the defence implications of urbanization trends.

Two broad questions dominate such speculation. The first concerns the type of urban development that has occurred, and is proceeding, especially in north-western Europe, and the degree to which this has significantly altered the defence characteristics of such areas – and, therefore, the military response already described in this chapter. A second and related question accepts that such changes have occurred and proceeds to examine the extent to which defence considerations should, and could, play a role in the shaping of such environments.

In practice, discussion on the first question revolves around the extent, nature and consequences of suburbanization. The urban environment discussed so far has been characterized by a dense plot-occupancy by multi-storey buildings, narrow streets and restricted open spaces; to which military operations have responded in the way described in the detailed example of the battle for the inner city of Groningen. The suburbanization of the western European population, which became especially marked in the decades after the Second World War, has resulted in an expansion of the cities to include extensive areas of low-density, one- or two-storey housing, wide areas of open space, broad access roads within such areas and high-speed motorways linking them with each other and with national transport networks. In essence, the problem posed by such areas is this: should they

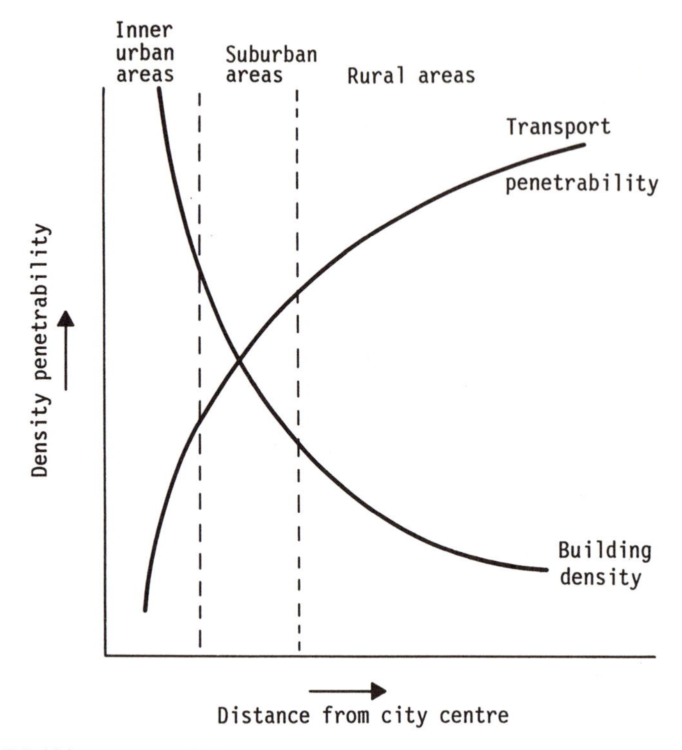

Figure 5.5 Urban penetrability and defence

be regarded as providing the defensive advantages of the urban battlefield listed above or the characteristics of accessibility, and penetrability of rural areas?

Figure 5.5 (adapted from O'Sullivan and Miller 1983) plots likely speeds of advance against distance from the city centre, making the assumption that these suburban characteristics increase in their incidence with greater distance from the urban centre. The 'attack velocity' (depending upon road density) and the 'defence velocity' (depending upon building density) can then be plotted. The 'trade-off' between accessibility (benefiting mechanized and armoured advance) and building density (favouring static defence) has a clear outcome in both purely rural and urban areas but remains uncertain in the intermediate suburban zone. The practical importance of this theoretical discussion is underlined by Figure 5.6. The assumptions upon which both NATO and Warsaw Pact forces based their scenarios of future conventional war in Europe depended substantially upon a clear distinction between the towns (used by NATO as defensive positions, and bypassed as far as possible by advancing armour) and rural areas (within which mobile battles will occur). The areas in the Federal Republic of Germany

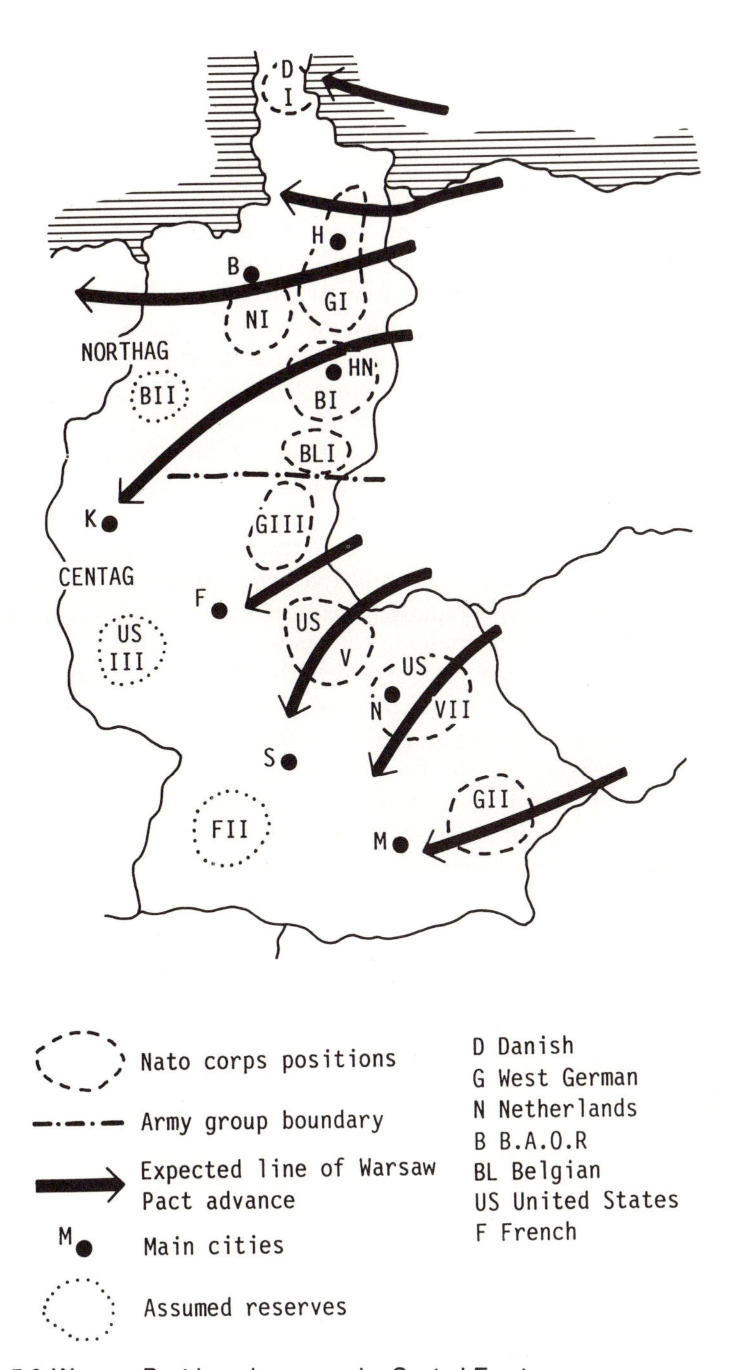

Figure 5.6 Warsaw Pact invasion scenario, Central Front

through which the major advances were expected between Braunschweig and Hanover, and between Fulda and Frankfurt, are precisely those which have been subjected in the last 20 years to the sort of suburbanization described above. The extent to which long-standing assumptions have been overtaken by a change in the character of the battlefield may clearly be critical in influencing the outcome of such a battle. Whether such predictions are realistic or not, they will be an influence upon contemporary military training, and raise questions about urban design and planning.

Although many of the urban battles referred to earlier undoubtely included suburban episodes, the nature of suburbs as battle terrain has rarely been explicitly raised. One of the few well-documented, division-scale battles to occur dominantly in a suburban and urban fringe area was at Arnhem in 1944. The airborne advance from the rural drop and landing zones of Wolfheze was blocked, with the exception of a single battalion that penetrated to the inner city and fought the 4-day action at the bridgehead. The resulting 9-day battle took place principally in the suburb of Oosterbeek. The environment included a relatively dense network of broad roads, which in this case benefited the defending German forces, who had access to both armour and more motor vehicles than the attacking airborne troops, and were thus able rapidly to reinforce their position in the western suburbs of the town. The building pattern was a mixture of large detached villas in extensive grounds (one of which, the Hartenstein, formed the core of the ultimate 'cauldron' British defence position) and two-storey detached and semi-detached brick houses with front and rear gardens bordered by small fences, hedges or walls. This, therefore, was a terrain easily accessible to infantry and providing defensible positions that offered some concealment but little protection, especially against armour.

Thus, contemporary trends in the design of cities pose sets of questions about the nature of the modern and future urban battlefield, and raise doubts about the value of such experience as can be gleaned from the limited number of historical examples of large-scale, conventional, urban conflict, few of which are less than 40 years old. It is possible to reverse the sequence of stimulus and response, and speculate about the possibility of planning the future battlefield. To what extent is it possible, or desirable, to include such defence considerations in the planning process, so that predicted battlefields are designed with their future defence requirements influencing the outcome of regional and urban planning decisions? Can cities be planned for future battles, in the same way as they have been for future insurgencies (see Chapter 4)? The answer most likely to be obtained, at least in countries with advanced public planning systems, is that such design is technically feasible but politically impossible. The existing statutory framework and technical and organizational expertise is used routinely to regulate, at the local level, the use of building materials, the siting of structures and open spaces and vistas, patterns of road access and the like, and at the wider regional and national level the planning of transport networks, interchanges and terminals, the locations of public utilities of military value and of defence installations themselves. Although these are precisely the characteristics that shape the urban battlefield, it is unlikely that decisions about them will take any account of their defence implications, unless quite special circumstances intervene. Planning for battle can be regarded in much the same way as planning for floods, earthquakes, and other natural disasters, or for the results of man-made accidents causing widespread danger or devastation. It is a problem of hazard perception and subsequent risk calculation. Viewed in this way, the chance of a city becoming a major battlefield in most parts of the world is too remote for much weight to be attached to preparation for it, in comparison with other more pressing and immediate considerations. The task of predicting what sort of battles will be fought over which cities is therefore, likely to be even more difficult than predicting most natural disasters. In addition, there is a difficulty (similar to that encountered in Chatper 6) with the planning reaction to the threat of aerial bombardment and nuclear annihilation. Preparation for fighting over the city requires a contemplation and acceptance of the possibilities of a scenario of urban disaster that would be unacceptable in local, democratically-responsive planning bodies.

Nevertheless, factors of internal security are increasingly playing a role in urban management and design. In Chapter 4 it was noted that planning and building regulations increasingly consider vandalism and crime, and that the deliberate shaping of urban morphological and functional patterns, to create a defensive environment favouring security forces over potential rioters or insurgents, has a long tradition. It is a relatively short step from this to considering battlefield-design factors in certain restricted categories of city, whose special circumstances favour such consideration. These would include: those with a recent history of political instability and military intervention (such as Beirut); those planted in part for defence purposes (such as the various Israeli West Bank settlements); or existing urban areas within the most likely future battle zones. The most obvious example in the last category remains large parts of north-west Europe, whose cities have in the past provided most of the world's urban battlefields, and whose future strategic importance, although uncertain, is unlikely to be diminished.

Chapter 6

The air defence of cities

ORIGINS

When viewed in terms of the technical development of projectile weapons, there is a clear logical line of progression that begins with the muscle-propelled stone or spear and ends with the rocket-delivered nuclear warhead. The bomber and the ballistic missile are thus only an extension of artillery, to which cities have for some 500 years been subjected. Such a technical continuum, however, conceals a distinctly new strategic role of cities, and a new set of reactions to that newly imposed role – introduced by the possibility of their attack from the air.

In 1849 the Austrian army, when somewhat unsuccessfully engaged in besieging Venice, tested the idea (first considered in the 1780s) of dropping explosive projectiles from balloons drifting across the city. This was correctly hailed at the time as the beginning of a new era in urban warfare, despite the failure to inflict either casualties or damage, or even hit the city with all but one bomb. Three conclusions were immediately drawn by attackers, defenders and neutral observers alike. First, a new front had been opened – the skies above the cities – in which the attackers had largely unrestricted and unchallenged freedom of movement and selection of targets, while the defenders were effectively defenceless. (Balloons had long been used for reconnaissance – as in the 1796 siege of Mainz and artillery spotting, but not in such a directly offensive role.) Second, the objective was clearly recognized as being the intimidation of the civilian population, removing at a stroke the centuries-long distinction between legitimate military, and immune non-military, targets, placing cities in the front line of future engagements and placing their civilian populations in as much danger as military personnel. Of course, bombardment intended to intimidate a civilian population was not new: Britain's incendiary rocket attack on Copenhagen in 1807 (Lindeberg 1974) is a textbook example of its deliberate and successful use, but it remained an option and was not an inevitable use of the weapon. Third, Radetsky's use of aerial bombardment was greeted with a near universal moral outcry, and there was much soul-searching among both civilians and professional soldiers about both the legitimacy of such methods and their efficiency.

Although it was to be almost another 100 years before aerial attack was technically capable of fulfilling the predictions made for it, these first reactions were sustained. Attack would prove incomparably easier than defence, cities became primary targets to be held hostage by the threat of terror, and a set of moral issues became intrinsically entangled with the use, preparation for use, and even defence against use, of such weapons. This was reflected in a series of international agreements, beginning with The Hague convention of 1907 – whose Article 25 outlawed 'the bombardment of undefended towns' but allowed attacks on military targets within such towns. This ambiguity was continued in the 'Hague rules' of 1922 which similarly forbade both 'terror bombing' and 'city bombing' but accepted 'accidental civilian casualties'.

This chapter is a recognition of the importance of this development for cities over the last 100 years, if only because of the blunt reality that the principal effect of the military function upon cities is that on occasion it destroys them, and even the threat of this possibility is both a defence function of cities and an omnipresent feature of urban living. The evolution of this situation over the last century will be related as an oscillation between theory and application, with the former being based on imaginative extrapolations from the latter. The defensive reactions of the city has similarly been dominated by the difficulties of choosing between a very limited range of alternatives. These experiences culminate in the reaction to the contemporary nuclear hostage role of cities.

THE EXPECTATIONS AND THE EXPERIENCE

From the dawning of the first possibility of attacking cities from the air up to the present, a series of debates have occupied both professional and popular attention. Although the terminology used has varied through time, and the arguments have been stated in a number of different ways, the essence has always been the choice between three sets of polarized alternatives. There has been the tactical versus strategic argument, which is about whether airpower is best used in conjunction with other arms on the battlefield or as an alternative, independent, offensive operation. Second, there has been the choice of objectives of attack – between the traditional, Clausewitzian goal of the destruction of the enemy's armed forces, or (more fundamentally but less directly) the destruction of the national will to continue to resist. This argument becomes entangled with the debate about the selection of precise or more diffuse targets: precision as against area bombing, with the latter being used as something of a euphemism for city bombing. In practice, much of the substance of these debates on the proper use of the air-attack capability against cities has been conducted at cross-purposes – perhaps intentionally, as moral constraints and technical considerations become intertwined. The history of the development of attack from the air reveals a continuing gap between what was expected of it and the technical possibilities it could achieve, as well as between the political attempts to control its use and its application by the military.

It is not too far-fetched to compare the similar features of two small-scale uses of air power against civilian settlements in the same country, Libya, although separated by 77 years of technical and moral debate and development. In 1911 the Italians made one of the earliest uses of airplanes to bomb the settlement of Ain Zara in their campaign against the Turks, and in 1986 United States land- and carrier-based planes attacked targets in Tripoli and Benghazi. On both occasions, the principal objective was to damage civilian morale and weaken its support of the ruling regime rather than destroy the military capability of the defenders. On both occasions, the attackers claimed (despite the stated objectives) to have carefully chosen and attacked selected targets, while the technical impossibility of achieving such precision resulted in 'unexpected' civilian casualties and, in both cases, internationally embarassing damage (to a hospital in 1911 and to a 'neutral' consulate in 1986). The anticipated consequences were political rather than military, while the actual results on both occasions were somewhat disappointing and arguably counter-productive.

One reason for the exaggeration of the anticipated effects of aerial attack upon civilian populations was its early adoption for imperial policing. This 'air control', as termed by the British, offered advantages to the new airforces eager both to justify their independent existence and to try out the possibilites of the new weapon in an environment that was both militarily and politically safe (since they ran few risks of effective counter-attack, or political objections if the targets were rebellious tribesmen in remote areas). To a hard-pressed treasury, the financial advantages of policing by relatively cheap aerial reprisals were obvious. The British in Somaliland and Iraq, the French and Spanish in Morocco, the Italians in Libya, all experimented with the technique. From the beginning, therefore, the air weapon was seen as a strategic arm to be used for area attack on civilian settlements in order to influence the will to fight (or to rebel), rather than as an extension of tactical battlefield artillery, whose objective, was the destruction of enemy forces. The current nuclear threat to cities was not, therefore, solely a product of Hiroshima and Nagasaki in the post-war world but its essentials were intrinsic to the use of airpower almost from its origins.

Although the Bulgarians bombed Adrianapole in 1912, it was the First World War which provided the opportunity for testing the potential of the weapon as a major part of conventional warfare against 'civilized' populations. However, this was in many ways a frustratingly incomplete experience for the prophets, who expected a demonstration of the science fiction prediction of 1913 that in a war between France and Germany both

Paris and Berlin would quickly be destroyed (Kennett 1982). The destruction of neither city was, of course, within the capability of the nascent air forces: the former was ineffectively attacked in 1914, while the latter was well beyond the range of existing aircraft or dirigibles throughout the war. Instead, airpower was used tactically in a ground-support role, and the idea of a strategic air offensive as an alternative to the stalemate of the trenches was neither technically possible nor seriously considered. There were air raids on cities which registered a number of historical 'firsts' – such as the French raid on Freiburg and the Royal Naval Air Service raids on the Zeppelin bases of Friedrichshaven and Tondern. Most notable was the sustained campaign of Zeppelins against London in early 1917 in which the 300 tons of bombs dropped resulted in 1,400 casualties. Partly in response to this, the Forty-first (Independent) Wing RFC was formed – to operate, as the name suggests, independently of ground forces in attacks on targets within Germany (Messenger 1976). In all of these, the ostensible targets were military, but such precision was impossible to attain, and the results actually intended were civilian unrest and the consequent political discomfort of their governments.

It was not until late in the war that airforces were sufficiently equipped to contemplate a strategic role, and military leaders were beginning to appreciate the potential. As Allied Supreme Commander Foch realized in 1918, 'the effects of massive bombing attacks would be almost incalculable'. The experience of the war as 'the early primitive beginnings of direct air attack on centres of population' (Slessor 1954), therefore tantalizingly suggested an unfulfilled strategic potential as an alternative to the long-drawn-out attrition of the Western Front, without much real evidence of its effectiveness.

The field was thus open for the prophets of airpower to argue their case almost unopposed. Professionals defending the independent existence of their new air services, such as Trenchard in Britain and Mitchell in the United States, were joined by the new military theorists, such as Douhet in Italy (whose influential book, *Command of the Air*, based on little practical experience, appeared in 1921) and Spaight in Britain (whose book *Air Power and the Cities*, was published in 1930). The arguments were simple, and readily understood by the general public and their political leaders. Future wars would be decided by a decisive, immediate, and (in this respect) humane, 'knock-out blow' against major cities, aimed at civilian morale and not at enemy forces, which would be ignored as largely irrelevant. As Spaight put it:

> It is the sovereign people who will war today, and it is their nerve and morale that must be broken. The great cities are convenient assemblages of the sovereign people: therefore smash the cities and you smash the will to war. (Spaight 1930: 230)

The accuracy of such bombing was calculated using the results of artillery fire, and casualties were extrapolated in terms of tonnage of explosives, which led Douhet to conclude that a fleet of 300 bombers in eight raids would inflict an intolerable four million deaths on a major European conurbation. Against such 'pure terror' there was believed to be no defence.

Both interceptor fighters and anti-aircraft fire had proved to be ineffective in the First World War. The former were too slow and had no means of direction on to the bombers, while the latter had a hit ratio of around one to every 6,000 rounds fired: in any event, the effects of both would be to encourage night bombing and higher altitudes, both of which removed any pretence at precision. The first rudimentary air-raid precautions were introduced in Britain in 1924, and in France and Germany in the course of the 1930s. Britain felt itself to be the most vulnerable of the major European powers because of its physical densities and distribution of population and productive capacity and also perhaps because of the psychological shock of the sudden withdrawal of the traditional projection of its island status. This led Prime Minister Baldwin to conclude in 1932 that:

> There is no power on earth that can protect the man in the street from being bombed. Whatever people may tell him, the bomber will always get through.
>
> (Baldwin 1932)

The only possible defence, it was argued, was counter-attack with the same weapon.

The few large-scale, practical demonstrations of these arguments between the wars – such as the Japanese bombing of Chinese cities (especially Canton) and the actions of the German Condor Legion during the Spanish Civil War (most notably the attack on Guernica in April 1937 which caused 1,500 deaths) – appeared to confirm them in full.

The first months of the Second World War were therefore something of an anti-climax. With the exception of the bombing of Warsaw in September 1939, the bomber fleets of Britain, France and Germany failed to deliver the expected 'knock-out blows'. They were both unwilling to invite retaliation against their own vulnerable cities and were technically incapable of delivering such a blow against a major adversary at that time. The German fear that their timetable of blitzkrieg was in danger of interruption during the 5-day campaign in The Netherlands (10–15 May 1940) led to the decision to systematically destroy the major Dutch cities from the air – commencing with Rotterdam and moving if needed successively through The Hague, Amsterdam and Utrecht. The urban-hostage strategy was completely successful, and even the demonstration bombing of Rotterdam was actually unnecessary (de Jong 1970).

The first major, strategic, air offensive – that of the Luftwaffe (principally against London), did not occur until August 1940 and continued until May

1941. Although some 57,000 tons of bombs were dropped, of which around two-thirds were on London – more than enough, according to the inter-war theorists, to destroy the entire city – it was clear from the length of the offensive that the single 'knock-out blow' did not occur. In addition, despite some early misgivings of the British government, widespread panic or dissaffection with the war did not materialize among Londoners, who perversely demonstrated that they could 'take it'.

This failure, however, was a result not only of the greater than expected resilience of the victims' morale during 'the blitz' but also that the 'kill rates' per ton of explosive were much lower than expected, and defence through radar-guided fighter interceptors and anti-aircraft barrages was more effective than anticipated. In addition, planned evacuation was at least partially implemented and air-raid precautions – first suggested in a local government circular in 1935 and implemented from 1937 – were accelerated. There was a degree of ambiguity and hesitation in official attitudes to urban shelter provision. Objections to providing deep public shelters were based on the diversion of resources needed elsewhere, but also on an almost complete ignorance of the physical and psychological effects of prolonged bombardment and crowding people together, which it was feared might encourage the very panic reactions that the enemy desired. Family shelters, the ubiquitous Andersons and Morrisons, of which two million of the former and half a million of the latter were issued after 1940, avoided most of these potential problems. The use of underground railway stations was a spontaneous reaction, reluctantly tolerated by the authorities, and purpose-built deep shelters were not available until 1943, when eight shelters with a theoretical maximum capacity of 64,000 places were provided along the London underground railway's Northern Line, largely in response to the pressure of public opinion.

The 1940–1 German air offensive, unlike that of the 1944–5 'V' rockets, was not ostensibly aimed at the destruction of cities and their inhabitants. The RAF campaign against Germany that began in 1942 was quite deliberately and openly so designed. The operational and political justifications for this strategy have formed the substance of a long-running, post-war, public debate which need not be repeated here. The political necessity of opening some 'second front' on the continent of Europe corresponded with the conviction within the RAF, most clearly articulated by Air Marshall Harris, that anything other than night-time area bombing was both very costly and largely ineffective. When the criterion of accuracy was the delivery of more than 60 per cent of the payload within 3 miles of the target (Kennett 1982), it is difficult to regard precision bombing as a realistic alternative. Whatever the justification, the objective of the RAF, and also the USAAF after the so-called 'point blank' agreement between them, was quite simply and openly to destroy cities – Churchill's 'de-housing of the German worker' (Whiting 1987). Success was measured in acres or square miles destroyed, and was to be achieved by a concentration in time and space so as reach a critical mass that would overwhelm the defenders. Lübeck (March 1942) was the first experiment, but the first real attempt to destroy a city in a single raid was Cologne (May 1942) when more than 1,000 bombers were brought over the city within two and a half hours. Hamburg was an even more spectacular demonstration of the technique (Middlebrook 1980). The city was selected for destruction in July 1943, and it was estimated that 10,000 tons of explosive was the critical quantity needed for this purpose. The campaign which took from 24–28 July, created the first fire-storm, killed around 45,000 citizens and levelled 13 square miles of the city. The technique was repeated for most of the next 2 years, most notably the February 1945 raids of the USAAF on Berlin and the RAF on Dresden. The resulting physical effects on the cities has been carefully mapped by Mellor (1978).

In the Pacific theatre, deliberate city annihilation was an American goal from the beginning of 1945, with again the rationale that a strategic air offensive against civilian morale was an alternative to otherwise costly engagements with enemy land forces. The Japanese cities had been protected by distance until the 1944 allied conquests in the south and central Pacific, but their method of construction, and use of wood, rendered them peculiarly vulnerable to fire. This vulnerability was deliberately and skilfully exploited, with precise calculations of the quantity, type and intensity of incendiary attack necessary for the generation of uncontrollable fires. City burning had become an applied science. The first experiment was Nagoya (3 January 1945) and the first real demonstration of success was Tokyo (9 March 1945), which resulted in 80,000–100,000 dead and 16 square miles destroyed. Subsequently, the USAAF worked conscientiously through the urban hierarchy, shifting targets to the medium-sized cities, since the larger ones no longer offered sufficient unburnt area to be profitable. By the summer of 1945, therefore, cities as small as Toyama (population 127,000) were attacked and 99.5 per cent of the built-up area destroyed – possibly a new record for completeness.

The dropping of the atomic bomb – on 6 August on Hiroshima and 9 August on Nagasaki – was thus a continuation of a long existing strategy and, in terms of the scale of casualties or urban destruction, these bombings were not in themselves particularly notable. The novelty was the sort of explosive, not the objective nor the immediate results as such.

The professional consensus view of the Second World War is that it was: 'above all the first air war; for the first time the belligerent countries themselves and their peoples became a primary object for attack' (Slessor 1954). There is no consensus, however, with Air Marshall Slessor's continuation:

> by its strategic bombardment of Germany the RAF (together with the American Air Force) made a significant, if not decisive contribution to

victory. As regards Japan the advocates of air power stated that this alone determined the outcome.

(Slessor 1954)

The perseverence, ingenuity and adaptability of the German and Japanese urban population was as evident as that of the British. German air-raid-precaution policy was, from the beginning, more dependent upon the provision of communal shelters than that of the British, and less so on long-term strategic evacuation of potential targets. Japanese policy was largely non-existent until late in the war. Cities as functioning residential and production units proved much more difficult to destroy than had been anticipated, and there is little evidence that the popular will to resist was effected in the way the inter-war prophets had predicted. It needed land armies to defeat Germany, and although in the Japanese case, 'for the first time in history, the enemy would be defeated without the previous defeat of his armed forces' (Lider 1985), it is at least arguable that it was the official realization that such a defeat was now inevitable, rather than a breakdown in civilian morale in the cities, that led to the Japanese surrender.

The end of the Second World War did not mark the end of the world's experience of the aerial bombardment of cities. A number of cases of relatively small-scale air attacks on urban targets, principally for their symbolic or propaganda value, have occurred sporadically in the last 40 years, either as an adjunct to the operations of conventional ground forces, as in the so-called 'war of the cities' during the Iran/Iraq war (1979–88), or as practised regularly by Israel, especially on Lebanese settlements, as part of anti-insurgency operations, reminiscent of the imperial air control of the inter-war period.

Although Pyonyang and other North Korean cities were bombed during the Korean War (see Halliday and Cumings 1984), the only example on a scale similar to the Second World War experience occurred in North Vietnam between 1964 and 1973. Here a strategic air offensive was again used, as a less costly alternative to ground attack and in order to exercise political leverage on an enemy government – either as retaliation for enemy attacks (as 'Operation Flaming Dart' in 1965) or to 'bomb them to the conference table' (as in 'Operation Rolling Thunder' 1965–8). Although a larger tonnage of bombs was delivered than on Second World War Germany and Japan combined (Fitzgerald 1973), urban targets – specifically Hanoi and Haiphong – were only occasionally attacked, most seriously in December 1972. Furthermore, in deference to world opinion, they were not subjected to systematic area or 'carpet' bombing. The interesting question was whether cities in less developed countries were more or less vulnerable to such attacks. It was hypothesized that their capacity for self-defence would be less but so also would the economic and technical complexity of the city, which would make disruption more difficult. In practice, it appeared that the first assumption was incorrect. In terms of the provision of both anti-aircraft and civilian shelter facilities, North Vietnam's 8,000 anti-aircraft guns, 200 SAM batteries, evacuation and dispersal of non-essential personnel, and individual street and workplace shelters, gave Hanoi the 'strongest air defence in the world' (Karnow 1985). The attempt to bomb North Vietnam 'back to the stone age' – as General Curtis LeMay, who had previously masterminded the 1944 air offensive against Japan, put it in 1968 – was both more costly (700 aircraft shot down) and less effective (around 100,000 civilian dead) than had been hoped. 'American investment in the bombing campaign was wildly disproportionate to the destruction it inflicted' (Karnow 1985).

It is not the business of this brief account of the historical experience to enter into either the moral arguments surrounding the use of the weapon, nor the professional discussion about the alternative uses of the forces employed. It is central to the theme of this book to understand how cities became primary targets of attack and the results of the application of these military theories. Hewitt (1983) has collated statistics, from various sources, of differing reliability, to describe the extent and distribution of what he terms 'city annihilation' for four victims of strategic air attack – namely, Japan, Germany, Britain and Italy (see Table 6.1).

Table 6.1 Comparative figures for area bombing of cities in the Second World War

	Britain	*Italy*	*Germany*	*Japan*
Number of cities	*c.*45	*c.*50	70	62
Area destroyed (square kilometres)	*c.*15	*c.*100	333	425
Per cent built-up area	3	25	39	*c.*50
Civilian deaths	60,595	56,796	*c.*600,000	>900,000

Source: Adapted from Hewitt (1983)

Against the figures shown in Table 6.1 must be set the argument that tactical precision-bombing in support of other forces could be as devastating, although usually over a more restricted area, as in Caen in 1944, and that the record number of urban civilian casualties is almost certainly held not by one of the air-attacked cities of Germany or Japan but by Leningrad – where the 'conventional' siege of September 1941 to January 1944 resulted in between 500,000 and 1,000,000 dead. In practice, cities often proved far less vulnerable to air attack as well as far easier to repair and eventually reconstruct than theorists had assumed, despite the increasing technical efficiency in city annihilation by the attackers. Even in the long list of German and Japanese cities compiled by Hewitt (1983), where more than 50 per cent of the built-up area had been levelled, much of the critical physical infrastructure of public utilities survived the destruction of the surface buildings. Even more important, so did the network of economic and social relationships that allowed the city to function. It is salutary to recall that no major city in the world has ever been permanently annihilated by aerial bombardment.

URBAN DEFENCE AGAINST NUCLEAR ATTACK

The experience of more than a century (related above) has culminated in the modern strategic hostage role of cities. The contemporary debates about the results of nuclear attack and the dilemmas these pose for the defence of urban populations have, in essence, many similarities with those previously discussed. The most significant differences stem from the technical advances of explosive design and delivery systems, which have permitted the return to credibility of the hypothesis of the single devastating knock-out blow. However, as in 1939, the hypothesis is based on theoretical extrapolations from less powerful weapons rather than empirical examples, which leaves both protagonists and opponents of the idea much room for debate (Brodie 1959). This has proved to be one of the few aspects of contemporary defence studies in which geographers have not only taken an interest but have had that interest recognized by the authorities. The significant role of geographers in the UK in questioning, from a number of different approaches, the officially propagated figures for the effects of nuclear attack is outlined especially in Openshaw *et al.* (1983, 1985). The three main possible defence strategies (namely, interception, evacuation and shelter) inspire as little confidence as they did in the inter-war period, with a counter-stroke policy being the most accepted defence. There has been little advance in the attempt to modify military and political exigency by moral imperatives through agreed controls, even though (unlike similar attempts before 1945) such a nuclear exchange threatens the devastation of neutral as well as combatant cities through radioactive fall-out.

This situation of 'mutually assured destruction' (or the acronym MAD) was openly articulated by the late 1960s, once both the United States and the Soviet Union (and their military allies) had both the nuclear stockpiles and so-called 'triad' of delivery systems – manned bombers plus land-based and submarine-based inter-continental ballistic missiles (ICBMs) – capable of surviving a first strike sufficient to claim a guaranteed 'overkill' upon each other. It is not clear, but usually assumed, that the destruction which is mutually assured is a high percentage (80–90 per cent) of the settlements of the combatants, together with unspecified world-wide effects (which are incidental to the theory). The mutuality is itself the guarantee of stability in this philosophy of deterrence. An implication of this is that attempts to defend targeted cities, in any of the by now traditional ways, is potentially dangerous – as it lessens the assurance of destruction, unbalances the nuclear stand-off, and thus presents one or other side with a so-called 'window of opportunity' for engaging in a 'winnable' exchange (Cox 1981).

Against this conclusion that all defence of cities against air attack is now impossible – provocative of the very pre-emptive attack it seeks to defend against, and an act of aggression in itself – must be set three developments that have become apparent since the MAD scenario was first conceptualized as the basis of superpower defence policy. First, so long as technical refinements in the destructive effectiveness of the air weapon keeps pace with the creation of defence facilities or plans, then the strategic balance is not upset – and this is very largely what has happened in the history of civil defence in the post-war world. Second, the reduction of all armed forces to a combination of small-scale conventional forces and massive nuclear strike capability – the 'trip-wire/immediate devastation' strategy – was supplemented later by the doctrine of 'flexible response', and subsequently in the mid-1970s by that of 'countervailing force' (Hackett 1982). In the latter doctrine an escalating scale of mutual destruction makes the defence of cities against smaller, more manageable attacks more possible and profitable. Third, the existence of the nuclear superpower stand-off has not removed the threat of less totally destructive conventional air attacks on cities outside the superpower homelands (as was demonstrated in Baghdad and Basra in 1991). Indeed such a stand-off, it can be argued, makes surrogate wars of this sort more likely.

Urban civil defence policies

Whatever the justification, the defence of the civilian population of cities has from time to time been the policy of most countries since the Second World War, whether committed or not to the two main alliance systems. Post-war civil-defence policies were initially based on extensions from Second World War experience, even in countries for whom that experience was second-hand. The organizational structures and personnel were an inheritance from those created in wartime, and the assumptions were related to the technology of that period. Despite the demonstration of the effects of United States atomic weapons in 1945 and of Soviet nuclear capacity in 1949, the assumptions of destructive capability during the course of the 1950s was not significantly different from the conventional experience of the 1940s. Given the size of the atomic weapons concerned and the lack of appreciation of the longer-term radiation effects, it is not surprising that civil defence was based on general assumptions of 'up to Hiroshima ×10', the existence of lengthy warning lead-in times, and a shelter duration of less than 1 month. Once it began to become obvious that all three assumptions were flawed, authorities were presented with policy choices, which can be summarized as:

1 Evacuation of the population of threatened target cities.
2 Shelter provision for populations in target areas.
3 Interception of incoming war-heads.
4 Inaction.

The third choice was regarded for most of the post-1945 period as a near technical impossibility. Policies in most cities have therefore oscillated uneasily between the first two. Public policy making has also had to contend with resistance to such policies, on the grounds that any potential defence both provokes instability in the unstable equilibrium of deterrence and encourages an unfounded confidence that such defence is possible (and thus nuclear exchange winnable). To this scepticism of both military 'hawks' and 'doves' can be added the problem of selectivity. If defence for all cannot reasonably be provided, how can the provision of defence for some be offered, by either the free market or at public expense, in an open society? It is not surprising, therefore, that policies for urban defence against nuclear attack in the last 40 years have been characterized by inconsistency, half-heartedness, partiality, incomplete information, secrecy and occasional laughable naïvety.

Although easier to consider under a series of separate headings, the alternatives were frequently pursued together, with varying emphasis at different times and places, and alternated with periods where inaction was dominant.

Evacuation policies

There are, as Ziegler (1985) has pointed out, three quite different sorts of evacuation policy, namely: strategic evacuation, where the populations of potential target areas are encouraged to move to safer areas long before an actual attack is launched; tactical evacuation, where existing plans are not activated until an attack is imminent or has already been launched; and post-attack evacuation of survivors. In practice, it is generally the second that is meant when such policies are discussed. The first needs much longer-term planning and would be politically difficult to enforce, as Britain discovered in 1939 when evacuated children and non-essential personnel soon drifted back to the cities when the attacks on them did not immediately materialize. Such policies could be merged with plans for the longer-term distribution of industry and settlement in countries large enough to attempt to relocate population and production centres beyond the range of attackers, as was a factor in Soviet development east of the Urals in the 1930s. The range and radius of destruction of modern nuclear warheads has rendered strategic evacuation pointless – except for military installations and command centres – in most countries, although China is believed to have operated such a policy in the 1960s (Campbell 1982).

Many countries have, from time to time, developed tactical evacuation plans, although the costs in terms of disruption have prevented these from being tested on any scale. In fact, it has been argued that the whole purpose of air attack was to cause such an evacuation of the cities:

> It is unnecessary that the cities be destroyed in the sense that every house be levelled with the ground. It will be sufficient to have the civilian population driven out so that they cannot carry on their usual vocations.
> (Mitchell 1930: 262)

In addition, evacuation may have increased the effectiveness of the attacks on German cities in 1942–4 by leaving empty and combustible buildings as well as fewer firefighters (Bak 1987). Plans could thus be seen as a means of regularizing and controlling otherwise chaotic and disruptive 'voluntary trekking', as most notably occurred nightly around a number of Italian cities during the Second World War. Britain had an evacuation plan for almost ten million inhabitants of the London region in the early 1960s, as did the United States for the major cities. Two considerations have caused the abandonment of such policies early in the 1970s. The first is the decrease in warning time available from the 2–4 hours assumed in the US plans to the 4 minutes expected in the 1970s. This is plainly insufficient time to activate such plans. The second is that in the NATO central front region, from the Baltic to the Alps, transport priority would be given to military traffic in such circumstances, and the problem is not how to evacuate the cities but how to prevent voluntary evacuation from impeding the movement of military traffic. A 'stay put' policy has therefore been adopted and would be enforced. Evacuation plans in most European countries are now limited to key government personnel and even these would find it difficult, in practice, to reach shelter within the time constraints. Only in countries which have considerable, available space, little competing military traffic and the possibility of avoiding becoming direct targets – such as Norway, Sweden, and Canada – are such policies still credible. Many other European countries – such as The Netherlands – have, and occasionally practice, urban evacuation as a response to potential limited local environmental or industrial catastrophes, which would conceivably be equally relevant in the event of small-scale nuclear attack.

After being effectively abandoned in the 1970s, some revival in evacuation proposals has occurred in some countries in the 1980s, perhaps because flexible-response strategies have suggested alternatives to an all-out nuclear exchange, or, more cynically, because shelter policies have proved too expensive and some alternative must be offered to the public. In particular, the Soviet Union is believed to have developed an impressively complex plan for relocating workforces and their dependants from urban to 'safe' sites within a system of duplicate factories linked by railways (RUSIDS 1982). The United States published a plan in 1978, refined in 1982, for the dispersal of the population from 319 urban areas – effectively all cities with more than 50,000 inhabitants – to designated reception areas, on the assumption of a limited nuclear exchange (Platt 1984). It has not been made clear how such an evacuation could be enforced in the face of: voluntary migration elsewhere; a refusal to move; or an already expressed reluctance of some host communities to accept refugees. Unlike the Russian proposals, the American ideas are very largely dependent on private road transport.

Shelter provision

The public provision, or the encouragement of the private acquisition, of air-raid shelters within the city has a number of distinct advantages over evacuation. Above all, because shelters are provided more or less where the population lives and works, they minimize disruption of production and transport, and can be used at very short notice. It is also conceivable that they instil a more tangible sense of security, or at least reassurance that preparations are being made, than evacuation plans which remain invisible until activated. It is not surprising, therefore, that one of the first reactions to aerial attack was to speculate about armoured roofs. The experiences of the Second World War and subsequent wars have tended to reinforce the reputation of shelter as a convenient and effective defence.

The most persistent disadvantage of shelter provision, however, has always been expense, and costs have risen steadily with improvements in the effectiveness of the attack, so that anti-nuclear shelters have now to provide protection not only against blast, but also against heat and atmospheric contamination for a long period of time (Tyrell 1983). A corollary of cost considerations is that adequate shelter provision for the whole population of potential target areas is enormously financially (and probably politically) prohibitively expensive: an estimate in the early 1980s for the UK was £60,000–80,000 million (RUSIDS 1982). Consequently, there has been a tendency in many countries – most eloquently described for the British case by Campbell, 1982) – for two separate shelter policies, to be pursued simultaneously. These are: the provision of a limited number of high-cost 'citadels' for the protection of key military and civilian government personnel; and the propagation of low-cost improvisation techniques offering minimal protection for most of the population (Home Office, 1981a). The purpose of such a dual policy is not so much the effective protection of the population of the target cities, but the survival of an ordered government structure, together with enough reassurance to the population to prevent a breakdown in morale in the crisis stage leading up to a nuclear exchange. Both objectives make these aspects of civil defence an important part of the MAD philosophy and its successors; for the assurance of mutual destruction would be weakened if no authority capable of retaliating survived a first strike, or if civilian populations anxious for their own survival undermined national resolve during the pre-nuclear exchange crisis.

In Britain, the provision of such 'nuclear citadels' was initially an inheritance from the Second World War, the most dramatic of which was the underground complex beneath the Whitehall ministries, and the eight

'Northern Line' bunkers mentioned on page 141. Various regional headquarters were subsequently constructed in the course of the 1950s outside the urban areas. In the United States, similar deep-shelter provision was constructed for military and governmental commands in various parts of the country, culminating in the NORAD command centre under the Cheyenne Mountains of Colorado in 1963. The realization of the increased explosive power of warheads, and the speed and accuracy of delivery systems targeted on to such citadels (whose locations were widely known), rendered them in practice largely obsolete soon after their construction (Phillips and Ross 1983).

The choice between the central provision of public shelters and allowing or encouraging the private provision of household shelters has been approached differently in various countries at various times. In the United States, the combination of the free-market philosophy and particular international crises has led to sporadic 'shelter panics', most notably after the Cuban missile crisis of 1962 (RUSIDS 1982). In Scandinavia and Switzerland, where there is a reasonable expectation of being spectators rather than belligerents in a nuclear exchange, there has been a combination of public investment and generally both subsidies and a planning requirement for private provision. As a result, Switzerland claims to have some shelter for 90 per cent of the population, mostly in purpose-built structures, as well as the world's largest shelter in the Sonnenberg Tunnel, near Lucerne. The use of enforceable building regulations to encourage provision in newly built housing as a result of the Federal law of 1962 (revised 1971), and the effects of this on urban planning, have been discussed by Heller (1975). Denmark can offer about 5 million shelter places, equally divided between public and private; Sweden has around 5.5 million and Norway can accommodate about 40 per cent of its population, half of this in public shelters.

Compared with these substantial investments, the rest of western Europe appears vacillating and half-hearted. West German cities, for long the most likely battlefield of a third world war, have been threatened by tactical as well as strategic nuclear attack. They had, however, no serious shelter policy until 1970 and by the end of the decade only some two million places. In The Netherlands, local authorities have long been empowered to construct public shelters financed by a general 'disaster budget', and some 185 have been built – mostly in combination with other public works such as metro stations and underground car parking. These, however, could accommodate only just over 2 per cent of the population, and their distribution is extremely uneven and appears to have little logic to it (Groen 1988). Until 1981, the UK had no public-shelter provision, the wartime inheritance having become obsolete in terms of structure and location, and no official encouragement for private provision. Reliance was largely placed on offering guidance to households for makeshift emergency protection when an attack is imminent (Home Office 1981b). The position in the Soviet Union can only be guessed at (Douglas 1983), but it is likely that it has proceeded through a series of policy changes not dissimilar to that of the United States. Shelter in the USSR was seen as a viable policy even in the cities in the 1960s, although the provision was provided entirely publicly, generally as part of house-building programmes. The stress on evacuation policies, from the middle 1970s until recently, left only a necessity for shelter provision in the cities for small groups of essential workers and for civilian control centres.

Interception

Although missiles are generally too fast for interception by protective, manned, fighter aircraft, they can themselves be used in a defensive interception role. Surface-to-air missile (SAM) batteries have proved very effective against attacks by manned bombers, as was demonstrated in North Vietnam. However, until recently, the interception of incoming missiles was regarded by most authorities as at best capable of providing a filter rather than an impenetrable screen. The replacement of bombers by unmanned missiles, and the subsequent development of multiple warheads, meant that a high proportion of losses could be accepted and 'overkill' still be achieved. The quantity and speed of incoming missiles made an anti-ballistic missile (ABM) screen for the defence of cities against such attacks (in the way anti-aircraft batteries were used in the Second World War) largely obsolete, although Moscow is believed to have attempted to develop such a protective screen in the 1960s and Riyadh and Tel Aviv were defended against 'Scud' missiles by ABM 'patriots' in 1991 with some success. The United States had rejected the idea of ABM defence by 1967, largely on the grounds of its failure to offer impenetrability, and agreement was reached with the Soviets in 1972 (amended in 1974) on the ABM treaty, which mutually called a halt to such developments (McNamara 1986).

In the early 1980s the idea returned in a different form, as a result of strategic and technical changes. The former included a disenchantment with the assumptions of the MAD doctrine, while the latter stemmed in part from the experiences of the space programme – especially satellite technology. The problem of detecting, identifying and destroying fast-moving missiles at a safe distance could be made easier by satellite observation, hence the popular name 'Star Wars' given to the American Strategic Defence Initiative (SDI). In fact, a number of different applications of this group of related ideas have been suggested (see Graham 1983). These include: a relatively cheap, point defence system for urban areas; a global ballistic-missile defence system for large continental areas (which filters incoming ICBMs in early trajectory); and, more tentatively, a space cruise option, where defensive missiles as well as the detection and activation systems are located in earth orbit.

The strategic objections to such defensive systems are twofold: (1) they may not be effective (in that a less than 100-per-cent kill rate would be compensated by an increase in the number of attacking warheads) and (2) if they were effective, or even believed to be so, their very development (or threat of development) would upset the balance of deterrence and thus provoke an inevitable pre-emptive strike. The paradox of such 'military metaphysics' (Hackett 1985) is that a defensive system can be an act of aggression, which is compounded by the equally curious conclusion that making the technology available to the potential enemy might increase rather than decrease security.

CONCLUSION

Although the rapid technical advances of the last 140 years that have brought air attack from balloons to Star Wars have an internal scientific logic, the strategic position that has been reached is historically unique and without precedent in past military developments. Two contradictory conclusions are possible. The city is now as defenceless as was Venice against Radetsky's bombs in 1849, with the difference that now a 'touch of a button' can guarantee the total annihilation of any city in the world regardless of its defensive preparations. Equally it can be argued that the nuclear stand-off, backed by overkill stockpiles and based on the logic of mutual deterrence, has rendered cities invulnerable through their very defencelessness, as any attack is rendered unthinkable by the automatic guaranteed total response. Living with such paradoxes may be a permanent feature of the modern world, the alternatives being either Armageddon, or (as appears a possibility at the beginning of the 1990s) a mutually agreed retreat back down the strategic scale of military development. This may render a massive nuclear exchange between superpowers obsolete, but low-technology 'Dresdens' remain possible and 'Hiroshimas' are likely to be within the capabilities of a proliferating number of countries. The hostage role of cities remains.

Chapter 7

The re-use of redundant defence systems

INTRODUCTION

Defence activities frequently have an impact upon the form and function of cities that long outlives the original military necessity that created them. The lines of long, obsolete fortifications, the fields of fire or inundation determined by superseded technologies and the buildings erected to accommodate long-departed armies and their supplies, can remain as visible features of the urban landscape, often shaping the internal morphological structure of the city, centuries after the defence functions which brought them into being have disappeared. Cities whose locations were selected on defence criteria, whose growth was caused and shaped by defence considerations, or whose economies, societies and urban self-image were formed in military service, generally continue in existence once that impetus has disappeared, but with a distinctive legacy, welcome or not, which is directly attributable to the previous defence functions.

Defence is not unique among urban functions in being in a constant state of evolution, and thus making changing demands upon the city. Most urban activities are continuously and relatively rapidly altering their functional requirements to be met by an urban form that can accommodate such changes only slowly and with difficulty. The argument of this chapter is that this inertia, although common to all urban activities – to a greater or lesser degree – is particularly marked with the defence function, which therefore results in past defence activities having an especially influential and long-standing impact upon the city.

The demands that defence makes on cities are characterized by their extreme volatility. The need for defence is generally either pressing and immediate (with the threat to security taking precedence over most other requirements) or is a minor consideration of the long term (which can be sacrificed for other, more immediate, needs). In practice, history has oscillated with great rapidity between these two conditions, so that in time of war or recognized immediate threat of it, enormous demands could be made of cities, their populations and industries, in the name of common defence, while in the interludes of peace, military establishments, procurements and investments would be severely curtailed. An obvious result of this tendency towards rapid change in the intensity of the defence demands made upon cities is the large-scale creation of subsequently abandoned defence systems. These may be relict forms such as defence works, rapidly and expensively created in answer to a threat and as instantly abandoned as that threat receded, or the buildings erected to house and equip mass armies that were quickly raised and as quickly disbanded. Equally, whole towns can find that their economies expanded through wartime demands are abruptly left in peacetime with an unviable economy and a lack of purpose, becoming in whole or in part redundant towns. The history of major military bases, garrison towns and defence-equipment manufacturing towns can usually be summarized as: short periods of economic boom, individual prosperity and national significance, separated by long periods of economic stagnation, relative poverty and national obscurity.

The relationship between technological change and changes in defence demands is far less obvious, and it is difficult to argue that rapid changes in military technology are the cause of rapid technical obsolescence in weapons systems and thus in strategy and tactics, resulting in turn in a high level of redundancy. Historically, the pace of technical change has (until very recently) been slow, reflecting no doubt the pace of change in society as a whole. One of Alexander's phalanxmen or Caesar's legionaries transported through a thousand years would have found familiar weapons, and tactics of attack and defence. Napoleonic methods of urban assault or defence were little different from those of 300 years earlier. The account in Chapter 2 of changing urban fortification techniques is typified by slowly evolving improvements in weapon and counter-weapon architecture rather than rapid and revolutionary change. Viewed more narrowly, however, it can be argued that the pace of technological change altered abruptly around the middle of the nineteenth century, when revolutionary developments in the natural sciences (especially chemistry and physics), in the applied sciences of metallurgy, ballistics, building materials, and in mechanical engineering, were all applied with increasing speed to military problems. This resulted in an acceleration in the amount, speed of adaption and impacts of change, beginning around 1850 and continuing at an ever increasing tempo to the present. In this respect, therefore, it can be argued that, over the last century, technical change in defence systems has been so rapid as to produce redundancy on an ever increasing scale, with works being rendered obsolete on completion and weapons developments being cancelled as outdated before they can be delivered.

In many cases, the very nature of defence artefacts aids their survival through the periods of neglect between their loss of military functions and their acquisition of new ones. Earthworks, masonry fortifications, concrete bunkers and emplacements of various sorts, were obviously designed with an intrinsic robustness. Medieval castles are more likely to have survived the ravages of time than medieval houses, and the only surviving evidence of Iron Age Belgic settlements is frequently the surrounding fortifications. Even deliberate demolition has proved in many cases to be more costly in effort than the value of the space acquired. The British countryside is still littered with the 'pill-boxes' that comprised the 1940 anti-invasion defence lines, as is the coast of north-west Europe with remnants of the German 'Atlantic Wall', principally because these are just too expensive to dismantle – for which future generations may well be grateful. It is noteworthy that most of the objects of military architecture that will be discussed later in terms of their importance to a valued and conserved defence heritage, survived through the centuries only because their robustness determined that no one was prepared to incur the considerable costs of destroying them.

The question of the relationship between technical change and the nature of the demands that defence makes upon cities has been raised, but there is also a parallel argument that relates military conservatism to the survival of military artefacts. There is a conventional wisdom that regards military leadership throughout history as instinctively conservative, with each new conflict being fought (at least initially) with the equipment, methods and approaches of the previous leadership. As has been argued earlier, although defence has not in fact been noticeably slower than society as a whole to adopt new technologies, military 'science' does suffer from the intrinsic problem of dealing with the subject of war – in which each occurrence is unique and can only be approached on the basis of past experience. There is, therefore, a tendency to resist the abandonment of existing weapon systems, defence works and methods, on the grounds that they may have value in a future conflict. For example, it is not all that surprising that the military in many countries objected to the dismantling of city walls at the end of the nineteenth century on the grounds that, although they were built many centuries earlier, they may have possessed a latent but unknown value in some unspecified future conflict. The reluctance of defence authorities to abandon works which, although old, may yet be effective and would certainly be difficult to reconstruct should an emergency require it, leads to a situation where defence properties can pass from active military use to conserved historical heritage, without passing through an intervening period of obsolescence, abandonment and neglect. In these senses, military conservatism can be credited with the survival of technically redundant features.

Figure 7.1 shows a typology of the main states of defence works and some possible lines of their evolution through time, which provides an outline structure for this, and the following, chapter. The general theme running through Figure 7.1 is the relationship in defence works of form to function, which is disrupted once the active defence use of these works is abandoned, and either new uses must be found for the existing forms or these are replaced by new structures to accommodate new uses. The distinction

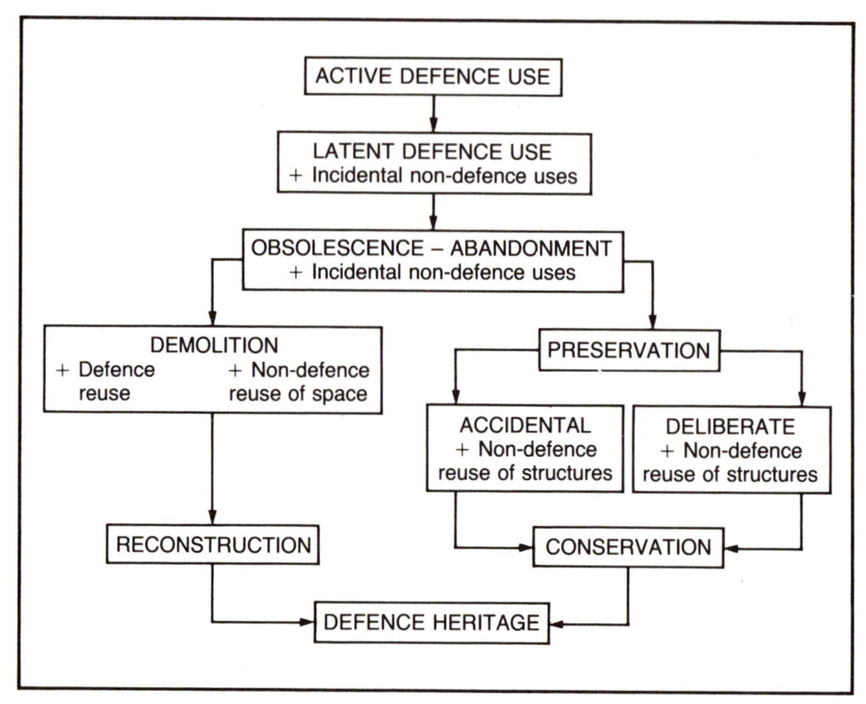

Figure 7.1 Re-use of redundant defence works: a schema

between active and latent use is often difficult to draw in practice. Military conservatism, or prudence, may lead to a very slow progression – from active use of works, through a latent use for works that might be resuscitated if needed, to the abandonment of works that are recognized as obsolete. However, incidental non-defence-related uses of such works (that is, uses for which they were not intended) can be expected to increase through this progression. Obsolete works, apart from being merely abandoned, can be demolished or maintained. On the one hand, if demolition is carried out, the space can clearly be re-used, either for new defence purposes or redesignated for quite different functions. On the other hand, the maintenance of the structures can be either a deliberate decision to preserve them for their own sake (in which case there is generally a necessity to find new ways of making use of them) or an accidental occurrence (in that new non-defence uses offer more advantages than demolition and redevelopment, and the existing structures are maintained to house them). In either event, the works remain in a state rendering them capable of being conserved.

Conservation is the deliberate ascribing of certain community values (whether aesthetic, historic or national-symbolic) to physical structures and places by governments or individuals, and the recognition of these values in designations which bestow at least protection and some access to financial subsidy for maintenance. It thus implies far more than preservation, whether deliberate or accidental. It is a further step from conservation to the idea of heritage, where the legacy of past structures are not being conserved for their intrinsic merit but are being presented as part of a definable product to a recognized market. The distinction between the use of existing preserved structures and the reconstruction of those previously in whole or in part demolished can, in practice, be a fine one. Nevertheless, such a distinction underlines a difference between conservational and heritage uses. Authenticity is important in conservation, whereas there is no necessary fundamental distinction between 'authentic' structures and reconstructed (or facsimile) ones in the use of past defence works as heritage. Once defence works have become heritage, a range of new issues are raised which will be considered in the following chapter.

THE CHRONOLOGY OF DISMANTLEMENT

The timing of the formal release of the fortification zone around European cities to civilian uses varied greatly between and within countries, and was often preceded by a long period of obsolescence. This was effective abandonment by the military authorities and thus the uses were informal, often illegal but generally tolerated (as described later in this chapter). This was only in part a result of an innate military conservatism. Existing works may be recognized as obsolete and largely abandoned while the defence authorities may still wish to retain the option of future modernization upon the existing space they occupy, in the knowledge that once such land is relinquished it would be extremely difficult to reassemble.

In considering timing, there is a clear distinction between an upgrading and relocation of a fortification system (as a result of either technical changes or the physical expansion of the town being protected) and its complete dismantling. The former could occur at almost any time in the 1,000 year history of the fortified European city, while the latter can usually be related to more general trends in national defence policies and to the location of the town within the continent. Replacement resulted in the release of continuous stretches of land within the new fortified city (often creating a morphological divide between urban districts), while abandonment usually resulted in the release of land on the urban periphery.

The chronology of the abandonment of urban fortifications is generally explainable in terms of the struggle between the military authorities concerned with the longer term and the broader defence picture and local interests reflecting more immediate demands for urban space. The resolution of these tensions was determined by the relative importance of considerations of national security against local economic pressures, in a particular place at a particular time. However, there was one additional factor (namely, the concept of the 'open city') which was accepted in the conventions of war from the seventeenth century, and increasingly formalized in the nineteenth. This (at least, in theory) spared a town from destruction if it was not fortified, thus providing an incentive to demolish fortifications which were believed to be ineffective. In the Low Countries, for example, those towns which found themselves outside the national defence lines created in the mid-nineteenth century had a particular incentive to demolish their now obsolete fortifications so as to be eligible for 'open city' status in the event of invasion. Furthermore, on a few occasions when local initiatives were not forthcoming, the central government actually made such dismantlement compulsory as a part of national-defence policy.

Although, in terms of military logic, a fortification system is an interdependent whole which should be either maintained or released in its entirety (which was generally the case), there are examples of piecemeal release. Many of the Dutch cities, for example, had succumbed to pressures for urban development and abandoned their clear fields of fire in the 1850s – some 20 years in advance of the fortifications themselves. For instance, during that period, sections of the fortifications were removed in Haarlem and Zutphen to allow railways entry to the town, thus weakening the remainder of the defence system.

In continental terms, there is the distinction between the European periphery (including the British Isles and most of Scandinavia, where the fortified city only exceptionally survived the Middle Ages) and the heartland of central Europe (where the more regular passage of armies encouraged its survival well into the modern world). In the strategically critical border areas – such as Alsace, Galicia or East Prussia – the fortified city survived into the twentieth century.

In both Britain and Scandinavia, seapower rendered walls unnecessary once centralized authority had been imposed, with the exception of the naval bases themselves (see Chapter 3), and, in the Scandinavian case, with the exception of the continental land and island bridge with the mainland. Stockholm demolished its walls in 1620 with the intention, never fulfilled, of replacing them. The experience of Napoleonic mobile warfare, together with the improvements in artillery technology described in Chapter 2, encouraged the development of national rather than local defence systems, and thus allowed many towns within wider 'national redoubts' to abandon their own local fortifications – for example, Utrecht (1824), Brussels (1830s) and Amsterdam (1837). Others that continued to occupy key positions within such systems had to wait much longer. The countries of Mediterranean Europe rarely developed such national systems, and Spanish and Italian cities generally abandoned their (already 100 years' obsolete) walls early in the nineteenth century. Those that survived – for example, Barcelona (1854), Madrid (1868) or Bologna (1902) – had little military value.

The Netherlands, with its long tradition of static positional defence, required that an individual town apply for permission to demolish, which was eventually extended even to frontier towns such as Maastricht (1867) and Groningen (1874). The Act of 1875, however, reversed this procedure, and all city authorities were given such general powers except for stated fortress towns. In France, Germany and Austria-Hungary, the fortress city maintained an important military function until much later – as witness the importance of Plevna (Bulgaria) in 1878, Metz and Belfort in 1870–1, or Przemysl in 1915–16. By 1918, however, even Germany had formally abandoned the concept and designation of the fortress town. The survival of city walls (such as those of Paris until the 1920s, or, less spectacularly, Weesp and Woudrichem in The Netherlands until the 1950s) was therefore a result of inertia rather than any lingering military value.

INCIDENTAL NON-DEFENCE USES

In the account of the development of the walled city in Chapter 2, it was clear that non-defence uses have been permitted, or at least were tolerated, alongside the military uses of defence works. The very nature of many urban fortifications that exist as a deterrent, whose success is determined by their not being put to the use for which they were constructed, means that their military purpose is latent rather than immediate, and that they will stand through long periods of peace as an open invitation to civilian uses. The shortage of space in the walled medieval city, exacerbated by those very walls, could also in part be mitigated by them. In addition, the temptation to build up to and against the wall itself (thus not only occupying available space but saving the labour and building material of a wall) was great. The complaint that this habit impeded efficient deployment and fields of fire could be ignored in times of peace, and one of the first defensive tasks when danger threatened in many medieval cities was widespread demolition of such excrescences. In order to be effectively defended, large parts of the town first had to be destroyed. The relatively dry covered spaces offered by towers, gatehouses, and even under parapets and walkways, was available for non-military storage or even residence, sometimes informally and sometimes – as in Utrecht (Sneep *et al.* 1982) – as a recompense in kind for the military services of the guilds that manned the watch. Less officially sanctioned was the traditional location of prostitution, in more than one Dutch city, in the wall arches (Zuydewijn 1988).

From the seventeenth to the nineteenth century, the increasing size of the land demands of military architecture vastly increased the opportunity for incidental civilian uses of both the works and their ever extending fields of fire. The ramparts, bastions, ravelins and the like (which, it must be remembered, often occupied a land area larger than the city they surrounded) provided the largest, and in many cases only, stretch of open space in the city. It is not surprising, therefore, that the fortifications generally fulfilled an important function as the main urban open-air recreation space. The twentieth-century segregation of military and civilian land uses would have been both untenable (on practical grounds) and unnecessary, when much of the defence was in the hands of a citizen militia. The ramparts provided the promenades and rides in summer and the frozen gracht was the city's winter ice-rink.

It would have been disastrous to the urban economy to attempt to sterilize the increasingly wider areas around the city devoted to the glacis, fields of fire, and existing or inundatable water defences. Until the coming of the railways, it was this zone that provided much of the city's food as well as providing the space, or access to water, needed for a wide range of industrial and commercial activities (for example, textile washing and drying in the towns of Flanders and Brabant). Such activities were generally permitted, with some attempt at regulation to prevent interference with military functions. Controls on the height and building materials of structures (so as to allow easy demolition) and even on the types of crops grown in fields of fire (so as to minimize cover) were often promulgated. The possibilities for an almost permanent state of conflict – over detail between the obvious, short-term, economic benefits to individuals and the often less obvious, longer-term security of the city – are unavoidable, and fill the council records of most fortified towns.

The deterrent nature of defence works, together with the military conservatism argued earlier, means that there is rarely a sharp distinction between active defence use and obsolescence. Fortification systems may be unmanned, neglected and in disrepair for long periods while remaining officially active and of latent value to defence. Such long periods of semi-obsolescence (which could extend over centuries) clearly encourages non-defence uses to become established, and renders the return to a state of military preparedness more difficult. For example, one of the main problems with the complex water-defence system of 'Fortress Holland' (described in Chapter 2) was its dependence in time of emergency on the instant inundation of large areas of farmland. This necessitated the deployment of substantial numbers of troops and military police to defend the sluice gates and enforce the inundations against a potentially resisting local population (Zuydewijn 1988).

THE RE-USE OF BUILDINGS

The obsolete defence buildings and structures may be reoccupied by civilian uses, thus saving the costs and possible inconvenience of demolition and rebuilding. The incidence of this depends upon a coincidence in requirements. There are numerous examples of what amounts to a long-standing tradition of individual opportunism – such as the tolerated convention in most European walled cities of taking advantage of the 'free wall' provided by disused fortifications for building. An extension of the 'free wall' is the 'free quarry', where citizens help themselves to a conveniently located supply of building materials. Throughout Europe, the bunkers, 'pill boxes' and gun emplacements of twentieth-century defence lines have been used for storage and agricultural outbuildings. In the cities, air-raid shelters (especially the mass-produced British, Anderson, outdoor shelters) have remained, to serve a generation as tool-sheds, rather than incur the costs of removal.

Finding suitable civilian uses for the larger buildings has generally proved more difficult. Depots and workshops have similar requirements, whether in military or civil use, but there are few equivalent civilian uses for barracks. Once the demand for communal housing has been satisfied, and the city has exhausted its relatively restricted needs for prisons, residential hostels and the like, then extensive and expensive conversion is required if barrack accommodation is to be rehabilitated for family housing, hotels or offices.

It is not difficult to find well-publicized, show-piece examples of successful conversion. The Puckpool battery at Plymouth has become luxury holiday accommodation and the Fredericks barracks in central Leeuwarden was converted to private apartments. More generally, the 'castle museum' is a ubiquitous combination of form and function in countless European towns. In The Netherlands alone, in the last 3 years, military buildings have been reoccupied as commercial offices (Lochem and Voorsteden), residential apartments (Renswoude), town halls (Ruurlo and Vorden), university teaching accommodation (Leiden) and a conference centre (Vaals) – see Stevens 1987. These examples, although often imaginative, are somewhat misleading – as they represent only a small proportion of the total available and, generally, the larger and more notable buildings. Rehabilitation, rather than demolition and purpose-built new structures, is exceptional, and usually only attempted if there is a compelling reason, supported by financial subsidies, to preserve the building itself. As the majority of military buildings were essentially utilitarian structures, this is seldom the case.

The successful transfer of buildings from military to civilian use depends upon a coincidence in both requirements and timing of release which is relatively rare. However, the extensive results of the existence of just such a set of coincidences has been described by Riley (1987) for the naval base of Portsmouth. Here the contraction in the land requirements of a major military base was a long process of piecemeal release (beginning around 1870 and continuing to the present), rather than a single sudden event. This contraction in military demand coincided with the growth in the demands of a number of private and public civil functions, most notably the rapid expansion of the polytechnic. In particular, the requirements of education were frequently sufficiently non-specific to be accommodated in barracks. In the case of catering and sports facilities, the requirements were intrinsically the same as those of the military. It should be added, however, that the occupation of military buildings by public institutions was more often than not a result of their lack of financial resources rather than their expressed preferences for such sites. Former barracks in the UK are currently being occupied by: a private school (Cambridge Barracks); city museum, archives and art gallery (Clarence Barracks); and by various polytechnic functions, including a staff sport and social club (Nuffield club), student-catering facilities (NAAFI) and a number of teaching and administrative departments (Milldam Barracks, Ravelin House).

The linear nature of many fortifications presents some opportunities for their direct re-use for civilian purposes. The use of the wall itself, and not just the space it occupies, as a route system survived in York and Canterbury to become later a tourist attraction. The water defences of continental cities may be usable for navigation – when they do not follow too circuitous a course around the artillery bastions and when they have sufficient depth. In the case of Groningen (see Figure 2.12) the southern grachten were used as the basis of a new canal (linking the existing waterways) and for new harbour basins; the northern section was retained for its landscape rather than navigation value. However, both the eastern and western sections were filled in to provide new building land, as – in this case and more generally – it was more advantageous to dismantle and redevelop the space than to re-use existing structures.

THE RE-USE OF SPACE

The most widespread and useful urban legacy from previous defence functions is the space occupied by demolished and abandoned defence works. The value of this inheritance is partly its sheer extent – which, in the case of the important post-seventeenth-century fortresses discussed in Chapter 2, included not only walls but the whole ensemble of ditches, water defences, glacis slopes, counterscarps and outlying redoubts and bastions, together with their fields of fire. In the case of the smaller frontier fortress towns, it is not surprising that the release of such defence land more than doubled the area of the town.

The location of such land in relation to the stage of urban development at the time of its release was often equally important. Some such locations were central – such as parade grounds, which were often large, centrally-located, open spaces with associated processional ways. This contributed much needed public open space in the heart of the city. The survival in many of Europe's cities of important central open spaces can be credited to their previous use as military parade grounds, whether London's Horse Guards' Parade, Viennas Hofburg or the Champs de Mars of many French provincial towns. Similarly, peripheral open spaces had often been preserved from development by their military use, and their abandonment offered dramatic opportunities for urban expansion. For example, until well into the nineteenth century, Great Yarmouth was a relatively compact town on the river

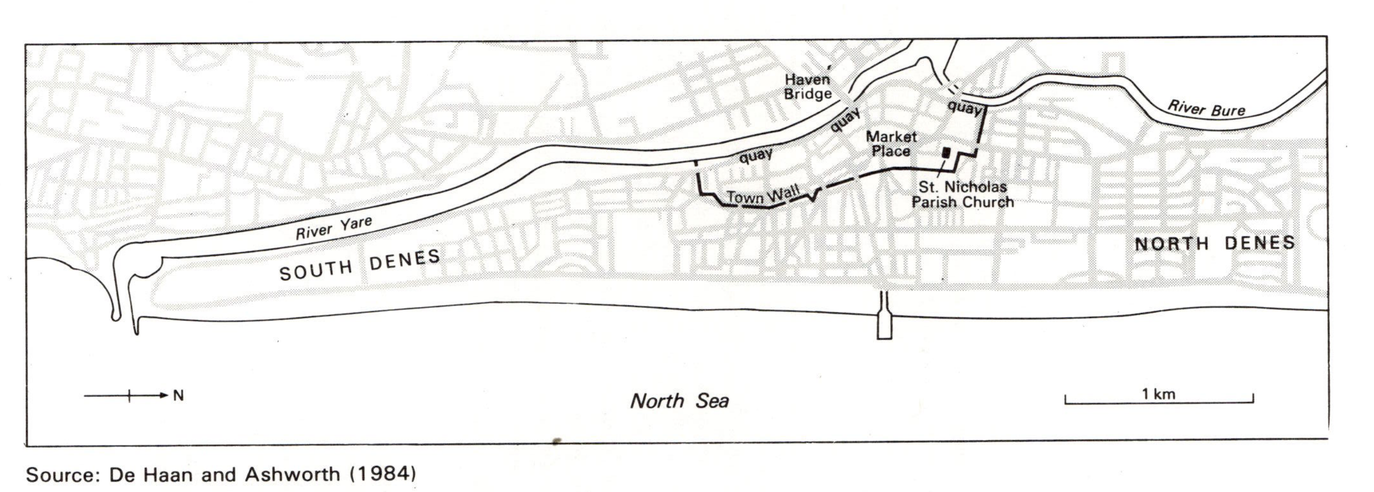

Source: De Haan and Ashworth (1984)

Figure 7.2 Use of defences in Great Yarmouth

Yare (see Figure 7.2), with its sizeable garrison making use of the 'Denes' – the sandy peninsula between the river and the North Sea – for military and recreational purposes. The development of the seaside resort after the middle of the century depended upon the release of this land. By contrast, some locations, such as those of barracks and depots were scattered through the city, providing infill possibilities of varying size. However, the most influential possibility was provided by the walls themselves because the land released formed a continuous ring, often but not invariably around the periphery of the towns.

The dismantling of the European urban fortification systems presented a unique opportunity for expansion. The reason for this uniqueness was partly because of the sheer quantity of land being released for development at one time, frequently together with its location in contiguous parcels, and partly that in most cities it was released to local authorities. In some countries the town walls were already owned by the town councils; in others ownership was transferred from military authorities to local governments on decommissioning. There were two main exceptions to this assumption of responsibility by town councils. The national government took a direct responsibility in two types of city: those large capital cities where the release was so extensive and symbolically significant that national effort was required; and, in contrast, those (generally, relatively small) towns which were major military fortresses – where the ministries of defence themselves undertook the dismantling and subsequent rehabilitation. In The Netherlands, for example, the fortifications of the city of Groningen were not freely released by the state to the municipality, although the city had largely maintained them since the seventeenth century (Boiten 1988). The town council had only participation rights in the replanning of its extensive fortification area in 1875, which was undertaken by agencies of the national government.

In most cases, however, it was town councils that were presented with a situation that had not previously occurred on such a scale, and of which, therefore, they had little experience. These councils were immediately confronted with two related sets of decisions. This unique windfall could either be used for public amenities or released to private developers. Furthermore, it could be developed piecemeal, by incorporation of the new space into neighbouring parcels as needed, or the contiguity could be maintained by planning for the development of the space as a whole.

Given the inexperience of most local politicians in large-scale, urban-development projects, the prevailing liberal sentiments in most of north-west Europe, and the dominance of many such councils by local commercial interests, it is perhaps surprising that so many towns did attempt to use the newly acquired defence space for civic purposes. In addition, there was a general lack of professional town-planning advice (especially with expertise in this specialized topic) available to town councils in the middle of the nineteenth century. The production of coherent schemes of urban design, many of which were highly imaginative, is in itself remarkable. Military engineers were skilled in construction, not destruction and rehabilitation. By the late 1870s in The Netherlands, however, the term 'dismantling engineer' was being used in development-plan submissions. Much depended, of course, on the particular pressures for development in the individual town at the moment of release. Nevertheless, the opportunity was seized by a number of major European cities to transform their structure and design – by grandiose, centrally-planned projects, generally supported by national governments. A detailed discussion of the reasons for this reaction – which has been acclaimed, as a result of its scale and comprehensiveness, as the birth of modern town planning (Hall 1986) – is beyond the scope of this book. It should be noted, however, that in a number of European cities (particularly capital cities, such as St Petersburg, Rome, Paris, Copenhagen, Brussels, Budapest and Vienna) there occurred a combination of purely civic pressures for new communal facilities (with national, imperial or dynastic needs) to enhance the residenzstadt as a visible symbol of the regime.

Not all cities reacted to their windfall gains of potential development space by producing large-scale, coherent, planned schemes. In fact the very spectacular nature of the transformation of the major capital cities may be misleading in suggesting that this was invariably the case. In some instances, private developers and individuals with properties adjoining the fortifications eagerly seized the opportunity of expansion from a compliant local authority. Materials from the defence works were sold, or just removed, for incorporation in new buildings or roads, and the resulting space developed as required – often by contiguous owners with or without legal title.

In other cases, land was released to towns which had remained content within their walls and which had little immediate need for expansion space. The small and relatively remote border fortress town of Bourtange in the north-east Netherlands, for example, was decommissioned in 1853, but the only practical use of the area so released was for it to revert to agriculture. The problem of converting walls and ditches into workable, and thus saleable, agricultural plots was solved by a parcellation system that included sections of both, so that the former could be used to fill the latter (see Figure 7.3).

Finally, there are cases where the fortification system was neither actively retained nor demolished, and the land was not re-used but simply abandoned. Many stretches of the Dutch urban grachten were just left as closed water systems which developed their own ecology as urban nature reserves (Sneep 1982). A more spectacular example is Constantinople, whose 7-kilometre land walls (which had rendered the city all but impregnable for a thousand years) lost their military significance after the city's capture by the Turks in 1453. Official inaction for 500 years and, more recently, the desire to preserve (without the financial resources to do more than prohibit development) resulted in a variety of informal uses of defence space. The land

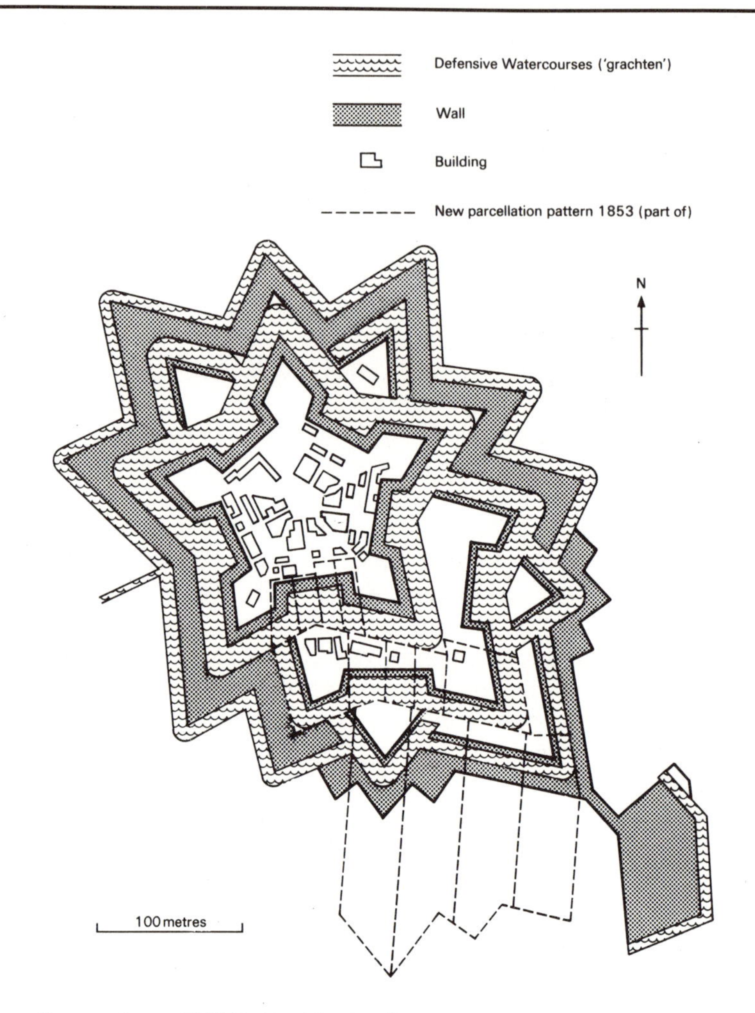

Source: Anon. (1982), Het Vesting Bourtange

Figure 7.3 Dismantling the fortress of Bourtange (The Netherlands)

between the walls (technically the 'peribolus') is used sporadically for squatter housing (including gypsy encampments) and at the Tokapi Gate for one of the city's most important, uncontrolled markets of more than 200 pitches. The space in front of the walls (the 'parateichion') and the surrounding ditch, when filled in, produced fertile, market-garden plots in a belt 100 metres wide by some 2 kilometres long, providing fruit and vegetables for the nearby inner-city population. The large variety of uses to which new space in cities could be put can be summarized under the following categories.

Further defence uses

Much of the land made available by the demolition of one military structure was promptly allocated to another. Being in the ownership of defence ministries, it is not surprising that this was frequently a first priority as a successor use. In the case of Portsmouth, analysed in detail by Riley (1987), no less than 70 per cent of the land area of the extensive fortification system reverted to other military uses, and three-quarters of the land area remained in crown ownership (see Table 7.1). This need not be ascribed only to the prudent maintenance of a land reserve by a conservative profession: the mid-nineteenth-century changes in technology that rendered the existing fortifications unnecessary equally made new, extensive, land-use demands. In Portsmouth the brick rubble from the old defences was promptly re-used for the new docks and basins needed for the new warships (Riley 1985). Similarly, the period of demolition generally coincided with the development of a new concern for the welfare of military personnel, which in turn led to a need for new barracks, social amenities and playing fields. The Portsmouth case can be regarded as having a number of unique features, not least its continuing importance to national defence and the dominance of this function over other competing land uses. A similar situation can be seen in other national fleet bases, such as Den Helder (see Chapter 3 and Figure 7.4), where all three of the forts are still in military use (housing a variety of residential, training and administrative functions) more than 100 years after their official abandonment as defensive positions. A residual military use, although on a much smaller scale, has also been identified in Dutch provincial towns (Sneep 1982), where the sites of obsolete fortifications have been re-used for barracks, depots and training areas.

Table 7.1 The re-use of land previously occupied by fortifications in Portsmouth, 1910

Function	*Hectares*	*%*
Military	51.7	70
Barracks	20.7	28
Residences	5.5	7
Recreation	15.1	21
Dockyard	10.4	14
Civilian	24.3	30
Recreation	7.0	9
Civic centre	4.3	6
Ecclesiastical	3.5	5
Commercial	3.5	5
Railway	3.3	4
School	0.7	1
Total	74.0	100

Source: Adapted from Riley (1987)

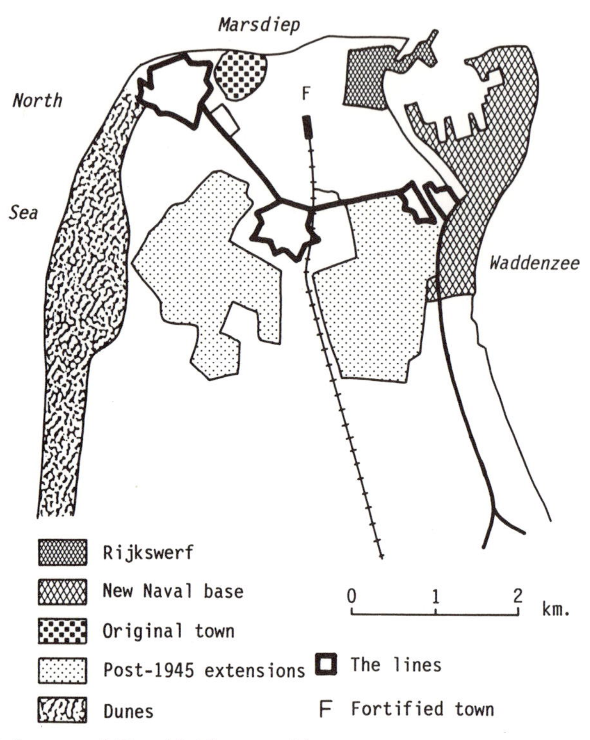

Figure 7.4 Defences of Den Helder naval base

In the middle of the nineteenth century, the release of many former fortification zones occurred after a period of considerable civic unrest and disorder in much of Europe, and at a time when more such disorder from the urban mob was feared by those governing them. Some of the planning reactions to this threat to internal security were discussed in Chapter 4, but it can be added here that many cities devoted some of the land made available by the dismantling of the fortifications to barracks intended for maintaining urban order. The newly available fortification zone often provided a morphological divide between the government buildings and richer residences of the old city and the working-class housing areas of the suburbs beyond. The argument was made in Vienna that this zone should be preserved as an internal defence zone (White 1984). Although this was not heeded, it was suitably placed for new barracks (such as the Caserne Verine in Paris, or the Kaiser Franz-Josef Kaserne in Vienna), which frequently survived well into the twentieth century. There was even an argument that the walls should be left for internal rather than external security.

The immediate re-use for defence, however, was frequently just a postponement of the ultimate release of land for civilian uses. In the major bases such as Portsmouth, in practice the result was that the defence authorities acted as a sort of filter, so that the large quantities of land made available by the sudden dismantling of fortification systems were only slowly released to civilian use – having passed through various residual military uses over a period as long as a century.

Civilian community uses

Despite the prevailing liberal ethos of most town councils, the importance of communal facilities can be ascribed to the timing of the release of land in many European cities. This corresponded with a growth in concern about congested slum-housing conditions and their consequences for public health and hygiene, together with contemporary ideas about the therapeutic value of open air and sunlight, and with the emergence of a certain civic pride in general. Parks, gardens and public promenades were therefore favoured uses for former bastions and glacis, which continued what had often been a long-standing, informal use of these areas for public recreation. One of the earliest, comprehensive plans was that of Haarlem in 1821, where the walls were converted to a circular park laid out in the 'English Style'. Cities such as Hamburg, Munster or Frankfurt created whole networks of new public parks around the inner city. To these uses were added new public utilities – such as schools, hospitals (see, for example, in Figure 2.12, the use of the whole eastern swathe of the city of Groningen for the regional teaching hospital) and even gas and water installations. Riley (1987) has noted a tendency for a continuation of institutional land use even after the transition from military to civilian ownership.

The linear nature of much of the land released, rendered it highly suitable for transport use, and the constrictions that the fortification systems had placed upon transport of various kinds encouraged the eager and early rectification of these transport constraints once the opportunity was created. Gates and entrances were quickly removed in most cities to improve road (and, in some cases, also railway) access into the inner city. In some cities (such as Haarlem and Zutphen) new railway stations were located on the glacis, while in Portsmouth the opportunity (long denied by the military) was taken in 1876 to extend the railway line to a new 'harbour station' for a ferry link with the Isle of Wight and its growing tourist traffic. The Dutch dependence upon both water defences and water transport enabled the former to be used for improvements to the latter, with the parts of the grachten being incorporated into canal extensions and even new inner-city harbours (see the ring of harbours in the Groningen case in Figure 2.12).

However, the most far-reaching impact upon the design and structure of cities was not the improvement of transport into the inner city but the development of new road patterns and associated land uses around it, following the line of the demolished circular defences. The most lasting

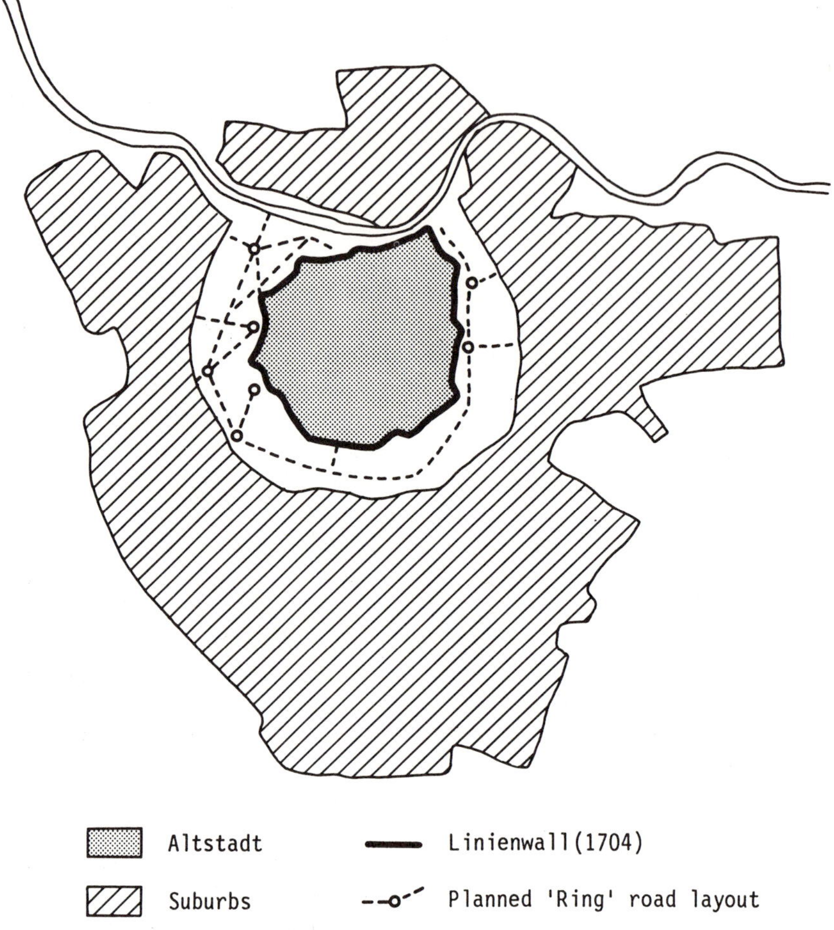

Figure 7.5 Defences of Vienna, 1850

tribute to the importance of this aspect of the re-use of defences is its contribution of two new words in the vocabulary of urban design. Boulevard (derived through a French corruption of the Dutch word bolwerk, or artillery bastion) was applied to the broad, imposing processional ways that were constructed on such sites, and esplanade (a military-engineering term for the open space in front of fortifications) was applied to the public, recreational, open spaces developed in such areas. Many spectacular examples exist, and the progress of their planning in this period has been well documented (see Hall 1986). Two of the best known and most comprehensive examples can be briefly mentioned, if only to stress the scale and importance of such developments.

Figure 7.5 shows the extent of the built-up area of Vienna in the early

1850s. The historic core is contained within its walls, at this time the Linienwall of 1704. Suburban growth outside this area, once the perceived threat of Turkish attack had receded, accelerated during the first half of the nineteenth century and had led to extensive housing development on both sides of the Danube, but an open space zone around the walls had been retained. By the middle of the nineteenth century, the eighteenth-century walls were clearly technically obsolete and had been replaced by a new system of earthworks and forts enclosing both city and suburbs. The opportunity for utilizing the inner defence zone in a grandiose new design was seized by the national government. Public ownership of the entire zone occurred at a time of emergence of a certain civic pride in general and, more specifically, a felt need for the assumption of new civic responsibilities which required new buildings for community use. In addition, this coincided with the need of the dynasty to assert the symbolic role of Vienna as an imperial capital ('Haupt und Residenzstadt Wien') of an increasingly fissiparous empire. The requirements laid down for the design competition specified a given number of civic buildings and road dimensions generous enough to amount to processional ways. These public functions were to be financed by prestigious residential development, expensive enough for sales to fund the rest of the plan. The winning design (approved in 1858 and executed, with few changes, over the next decade) included two theatres, three museums, a concert hall, law courts, a city hall, university and parliament buildings, as well as 590 dwellings, all of which were high priced. These were aligned along the system of roads and parks that formed The Ring.

The Vienna case is neither unique nor even the most extensive example in Europe. In other cities, however, the proportion of land devoted to public uses, and especially public open space, was generally much less. In Cologne, for example, the released land was used for residential development, with only a niggardly provision for a new ring road. In the case of Madrid (described by White 1984) the expansion on to what must have been one of the most extensive fortification zones around a major European city, after its dismantling in 1868, was again principally for expensive housing. The former fortification zone is recognizable not so much from its land uses as from the distinct differences in the morphological patterns of the three resulting zones (namely, the old walled city, the planned street pattern of the ensanche and the unplanned suburbs beyond).

The planning of the European capitals were reproduced on a smaller scale in many provincial towns which shed their fortress functions at this time. For example, in Arnhem (for long a defended crossing of the Nederrijn, with its fortress controlling an eastern entry to the western Netherlands), the dismantling of the walls released land for a double semi-circular boulevard around the inner city (see Figure 7.6). Apart from creating a new traffic circulation system (in effect, an inner ring road), the central reservation provided space for a variety of public uses, including parks and a concert hall

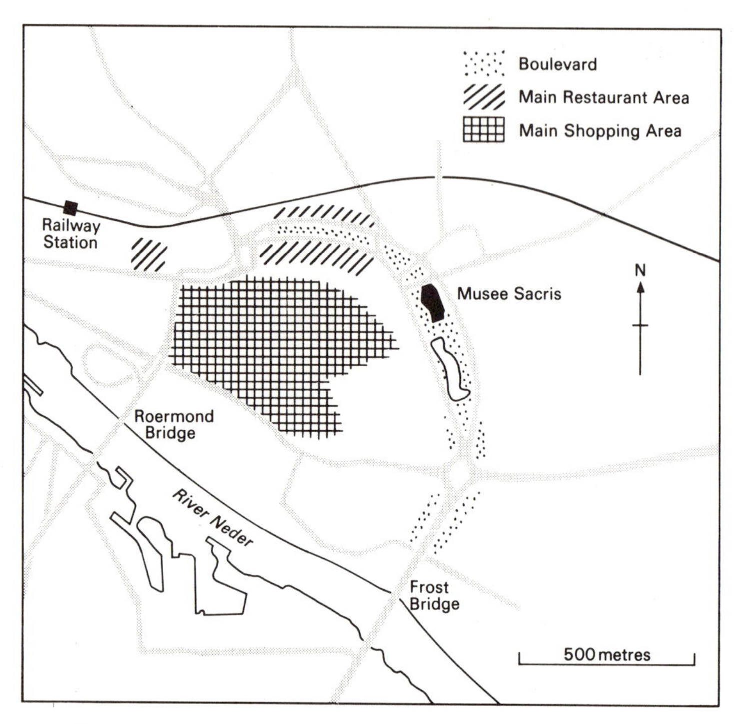

Figure 7.6 Re-use of defences of Arnhem

(the Musée Sacris). The land directly bordering these boulevards was developed originally as high-status housing for richer citizens – now able to move out of the cramped inner city into 3 to 4-storey prestigious housing overlooking the new landscaped boulevards. Beyond these, the land of the former glacis was developed over the course of the next 50 years for more modest housing. In the course of the twentieth century, the boulevard houses and apartments (often too large and expensive for family housing), especially as richer citizens migrated to the suburbs, became increasingly converted into commercial premises. The smaller buildings along the eastern boulevards became offices (especially for financial and legal services) while the western end became part of a restaurant and entertainment district, associated with the bus and railway station. The glacis housing developments experienced a physical and social deterioration, and now form a ring of low-income housing which includes the major concentration of ethnic minorities and the principal prostitution district (Ashworth *et al.* 1986), and is the subject of active urban renewal programmes. The line of the fortifications thus provides a distinctive morphological discontinuity in the contemporary town plan, and also accentuates the functional and social divisions of the city.

The dismantling of the northern fortress town of Groningen illustrates the difficulties of choosing between the two main types of development plan popular in the last quarter of the century. The 1875 development plan (submitted soon after demolition was permitted under the General Act of 1874) was a self-conscious, scaled-down copy of the Vienna Ring with noble vistas and spacious boulevards lined by expensive villas. This plan was executed on the southern flank, but disappointment with the absence of public open space led to the northern ramparts being laid out as public parks in 1880, with subsequent housing on the southern flank being on a more modest scale (Boiten 1988).

A national and municipal planning response to the release of the fortifications for civilian use was by no means confined to the major capital cities of Europe. Satoh (1986) has traced the conversion of the fortification zone of the many medium-sized Japanese castle towns to civilian uses during the Meiji restoration in the last quarter of the nineteenth century. The walls, water defences and flanking open space that separated the military/administrative districts from the newer residential and commercial districts provided sites for public parks and gardens, as well as space for new transport technologies, especially the railway.

Chapter 8

Defence as heritage

INTRODUCTION

The use of the term heritage – when applied to past military works, equipment or even spaces where past military events have occurred – has a number of immediate, important implications, and it should not be used, as it frequently is, as a synonym for conservation. While conservation draws its justification from the qualities of the conserved object itself, the concept of heritage implies both a legacy and a legatee. The first necessitates selection of what is to be treated as heritage from the existing stock of preserved artefacts, while the second needs a defined and targeted market. Both of these implications require deliberate decisions on the part of those selecting, packaging, presenting and marketing artefacts and associations. Heritage, therefore, does not just occur as a fortuitous survival from the past: it is created in the present, to satisfy the current needs of contemporary consumers, using objects and associations from the past as its basic raw materials (Ashworth 1988). A consequence of these implications is that questions about the product ('what is to be selected, on what criteria and for what purposes?') and about the market ('whose heritage?') assume a central importance in the shaping of what is to be regarded as heritage.

This is the case for all uses of the past and its surviving objects, but it is arguable that defence history in particular is a provider of an especially large, if not dominant, proportion of all such historic legacies, and thus defence heritage assumes a particular importance.

This may be explained in part by the tendency of defence artefacts to survive better than most the ravages of time (as discussed earlier), which has resulted in a more than proportional presence of defence-related objects among the visible relics of the past. However, much defence heritage relies not only on such visible objects but upon a miasma of invisible associations with military events with which such objects are, or can be, endowed. Indeed, there are many examples of places (such as the sites of battles) where there are no visible relics, yet such sites are indisputably part of the defence heritage. A more satisfactory explanation of the importance of defence to heritage as a whole is quite simply that organized physical conflict between people exercises a distinctive, widespread and extremely powerful emotional appeal. The dominance of war in children's play, in the output of books, magazines, films and television programmes for popular or critical consumption, and in the membership of hobby associations, all illustrate an obsessive and pervasive interest in this topic. Holmes has noted that: 'The popularity of military history is such that it has assumed the proportions of a minor industry' (Holmes 1985: 4). If places are regarded as 'the centre of individually felt values and meanings, or as a locality of emotional attachment and felt significance' (Pred 1984: 279), as the 'humanist' geographers have maintained, then, clearly, places and objects associated with defence are likely to arouse special attention and feeling from individuals. Equally, for the community as a whole, it has long been argued that places are receptacles of cultural values and acquire a sacred quality as symbols of such values (Firey 1945; Lowenthal 1975). Surviving defence works therefore become the recipients of these individual and social attributes.

These arguments offer explanations of the relative importance of defence as a part of heritage as a whole, but do not answer the questions raised earlier about what is to be regarded as heritage and the related question of whose inheritance is to be considered. The mere physical survival of artefacts, or the universally strong human emotions associated with the objects and places of conflict, are not sufficient to determine the selection, presentation or marketing of past defence works.

To assert that the writing of history was a continuous re-interpretation of the past by each generation according to contemporary values would surprise few historians (Horne 1984). This idea can be extended to encompass the conservation of historic artefacts and their presentation as heritage, by asserting that heritage planning is a conscious use of a selection of allusions to the past in order to satisfy the current needs of a contemporary market, or to propagate particular contemporary ideas through appeals to a selection of values evoked by past objects and occurrences. Such a description of much conservation planning, including the conservation of military artefacts, would be refuted by many practitioners, who would point out that age and intact survival were the principal criteria of selection, and that, in practice, there is little room for the exercise of other values (except occasionally the aesthetic) in deciding what is to be conserved. In most countries, the recent history of urban conservation as a rescue operation salvaging a few broadly acclaimed buildings, or areas, in the face of powerful development pressures (Kain 1981) renders irrelevant much of any argument over the values behind selection decisions. Once a heritage approach is adopted, however, and the past is packaged for consumers, there is much more freedom of action in the selection of both the past and the package. Heritage is therefore not an objective revelation of aspects of a fixed historical truth but the reconstruction of an interpretation selected on the basis of subjective criteria from numerous possible pasts. Such a reconstruction can be achieved with the help of: physical objects and structures that have survived from the past; place associations with the past; or, equally, modern animations, sound and lighting effects, theatricals or facsimile building.

Such a definition of heritage risks being regarded as obvious by those involved in marketing, or outrageous by conservation planners and architects. In understanding the use of defence works as heritage, however, it is essential that they are placed within a wider contemporary social context. Discussion on this aspect tends to focus around two issues: authenticity (generally resolved into a conflict between perceived historical truth and historical mythology) and ideology (i.e. the harnessing of the interest and feelings evoked by history to legitimate a particular view of contemporary society).

AUTHENTICITY

Once the surviving artefacts of the past are packaged for consumption and interpreted as heritage, the question of authenticity commonly becomes a central area of controversy. This may take the form of different interpretations of the aim of conservation: whether it is the revelation of a fixed truth or the provision of what the market expects; or within the conservation lobby between different versions of such veracity. The premises and content of most of these debates has little relevance to the practical question of the effectiveness of using heritage as a new function of redundant defence works.

As noted earlier, preservation is necessarily selective, and is based upon criteria whose effects have a large element of randomness. The chances of survival are higher for particular building materials, for particular types of defence equipment or for works in towns and regions not subsequently under redevelopment pressures. There is thus an exaggerated emphasis: upon urban defensive fortifications; upon static rather than mobile warfare in general; upon towns whose brief periods of historical importance to defence have been followed by long-term, economic stagnation or decline; and upon urban defence works in peripheral rather than core areas. Distortion through selective survival already exists before the conservation process begins its own series of selection processes from among the relict forms.

Conservation involves passing through some or all of a progressive series of actions (including protection, maintenance, repair, restoration and reconstruction). All these involve deliberate choice exercised by the responsible agencies on the basis of criteria. The extent and effectiveness of protection depends upon the balance struck between continuity and change. A number of biases are evident in what is protected and in what quantity, and – given that in most countries around 5 per cent of the stock of conserved buildings are destroyed annually (Dale 1982) – what remains protected. It is likely that the spectacular, large and unusual are preferred over the mundane, small and commonplace. In spatial terms, the differences in degree of protection offered between countries (see Dobby's 1978, comparative account of national conservation efforts), and among cities within countries, is enormous, reflecting the distribution of the resources and will to conserve rather than the intrinsic importance of the artefacts themselves, and giving an even less accurate reflection of the history of defence.

Similar biases can be detected further along the spectrum of conservation processes. Protection implies maintenance and repair, and restoration of what cannot be repaired. Two further difficulties arise here. First, there is no clear boundary between repair and reconstruction. It is a short step from repairing an existing city wall, replacing missing stones and walkways, and reconstructing stretches that have completely disappeared. Once those steps have been taken, the construction of facsimiles in the style of the past is an attractive means of compensating for the random results of preservation. Thus, as a result of the conservation process itself, many castles, city walls and other urban fortifications are more reconstruction than relic survival.

The second and, ultimately, related difficulty stems from the implication that restoration is a process of returning a structure to an original or authentic condition. Many urban defence structures are the result of a long process of adaptive re-use (as described in Chapter 2), which raises the problem of the selection of one past state, from among many, as the ideal to be restored. In practice, that choice becomes little more than a question of prevailing taste and fashion as to the preferred period, such as the eighteenth-century's preference for the 'classical' over the 'gothic', which encouraged the discovery and conservation of much of western Europe's Roman military heritage but ignored, and even removed, much of the medieval. Nineteenth-century romanticism largely reversed these priorities, and even led to the restoration of a past that owed more to poetic imagination than historical reality. The 'fairy tale' medievalism popularized by Pugin, Scott and Viollet-le-Duc resulted in Wales' Castell Coch, and numerous Rhineland castles and whole towns (of which, Carcasonne's 'cité' is the best known) that illustrate the dilemma of all restoration – namely, what is the authentic state to which it should be restored, or even whose idea of authenticity should provide the goal?

The above arguments derive from preservation and conservation, in both of which the object is the central concern, but once the use of such objects as heritage becomes important a whole range of new potential difficulties arise with the concept of authenticity. The biases and subjectivity already considered are compounded by those of the market. In practice, a heritage attraction is a combination of two elements: the 'site' (i.e. the intrinsic qualities of the place and its associations) and the 'marker' (i.e. the deliberate indication of such qualities to the consumer) – see MacCannell 1976. The necessary link between the conserved artefact and the user is provided by the necessary intermediary of the 'marker', which may be on-site or previously acquired information. The result is what MacCannell terms the 'sacralisation' of places through a process of 'enshrinement'. This process is cumulative, as such site marking is reinforced by use.

It is clear, therefore, that authenticity has little meaning divorced from its purposeful context. In terms of heritage, visitors 'collect' the sites that have been marked rather than those defined by any intrinsic criteria. The selection of heritage (from the stock of preserved possibilities) and its interpretation is a contemporary process which has little to do with the accurate revelation of an authentic past through its relics. The purposes that motivate conservation as 'the necessary myth' (Goodey and Ophir 1982) may be little more than vaguely articulated professional guidelines justified in terms of conventional wisdom, or they may be sufficiently coherent and logical to be dignified with the term ideology.

IDEOLOGY AND HERITAGE

The simplest explanation of the interest in heritage is curiosity about the origins of the present. In turn, such an interest in the past can be viewed as contributing the stability of continuity to an unstable and uncertain present (Ford 1978). The use of tradition to provide this sense of stability to the existing political or social order is so widespread as to form an almost universal function of the study of history and its relics. An extension of curiosity to an obsessive interest in a past which is seen as preferable to the present results in 'nostalgia', a word that means more than just a romantically tinted view of an idealized past, but literally a painful longing to return to it. The marketing of nostalgia through heritage can then be seen as escapism from an unattractive present and an undesirable future into a previous golden age. This escapism (Hewison 1987), or, more strongly put, 'cultural necrophilia' (Davies 1987) can be used as an instrument of political policy by governments who deflect a desire to change present undesirable conditions into nostalgia for a past (Hardy 1988).

Whether these uses of the survivals from the past (as tradition or escapism) amount to ideology, in the sense of a coherent political philosophy, is open to discussion. However, some of the more commonly encountered approaches – which may amount to ideologies – found in presentations of defence heritage are suggested below. Such a list is not exhaustive, nor is it unusual for a single presentation to contain elements from numerous approaches.

A nationalist approach

A distinctly ideological use of defence heritage is its consistent use to support a particular idea of state, which can be called nationalism where it is used to legitimate the nation state. Despite some attempts at finding a continental-scale replacement to the nation state in western Europe, nationalism remains the world's most widespread state-forming philosophy. Most such nationalisms require the mythology of a founding, armed struggle against a repressive folk enemy, from which crucible of fire emerges the national character and national values, which in turn must be defended 'against the envy of less happier lands'. The possible uses of defence heritage to support such national mythology are obvious. In the United States there are the relics and sites of the 'revolutionary war' and the civil war for the maintenance of the 'union'. In Canada, to refer to the other side of the same events, there is the war of 1812–15 against the United States. In Europe, the Spanish have the 'reconquista', the Dutch the Eighty Years War, the Belgians the 1830 war against the Dutch, the French the Hundred Years War, the Balkans the independence struggle against the Turks, and all can share in whatever national glory can be found in two world wars fought out across the continent. Meanwhile, there are few countries outside Europe which cannot find an independence struggle against a colonial oppressor upon which to base their self-esteem. Those which cannot, having achieved sovereign statehood peacefully, search uncomfortably among strikes, riots and skirmishes for the nearest equivalent – such as Australia's 'Eureka stockade' or romantic 'bushrangers', and Canada's Meti 'rebellion'.

The national myth will determine not only which defence works and associations will be incorporated into heritage but how that heritage is presented, so that the chosen central values of the state and the qualities of its citizens are substantiated by the chosen historical episodes associated with the objects and places. The converse is, of course, equally true. Objects and sites that recall the 'wrong' history will tend to be ignored. For example, American revolutionary war 'loyalists' play only the part of stock Tory villain in the United States but become the central heroes in the restoration of the Canadian heritage towns of Niagara-on-the-Lake and Queenston. The Dutch city where this is being written, Groningen, has suffered three important sieges in its long history, two of which are commemorated by physical memorials and public holidays – namely, 1672 (when the city was successfully held for the Dutch Republic against a German invader) and 1944 (when it was liberated from the Germans after siege by the Canadians). The third occasion is neither celebrated nor commemorated, as it conflicts with the 'national' idea: 1594, when the city, which was held for the Habsburgs, was besieged, taken and coerced into the Dutch Union (Schuitema Meijer 1974).

A particular, detailed example of the use of heritage in this way is provided by recent developments in the city of Portsmouth, and the commentary upon these by Bradbeer and Moon (1987). The importance of the defence function in the naval dockyard town of Portsmouth, and the survival of much of the defence works, was discussed in Chapter 3. Some use of these as heritage has long been made, especially HMS *Victory* and its associations with Admiral Lord Nelson and the Royal Navy. More recently, however, the city council – in partnership with various private organizations – has initiated a major series of heritage projects, including two further conserved ships (*Mary Rose* and *Warrior*), and a number of shore-based museums, under the overall marketing slogan of 'Portsmouth – Flagship of Maritime England'. These, together with similar heritage projects in South Hampshire and the Isle of Wight, form a regional product entitled, 'Defence of the Realm'. The ostensible justification for these developments is principally economic, as the dockyard town searches for a replacement for its declining staple activity of servicing the fleet.

However, Bradbeer and Moon (1987) argue that the choice of exhibits and their method of presentation is clearly: nationalist (in its account of the unremitting success of British arms); militarist (in its stress on the success of resorting to force); and imperialist (in its one-sided view of the impacts of the role of British defence forces). They further suggest that if heritage is seen as an instrument of social reproduction (i.e. the way in which the prevailing power structures in society are maintained), then Portsmouth's traditional role, as they see it, of subservience to the military is continued in this way. The philosophy of service to the fleet and wider national or imperial defence needs is continued by this particular philosophy of defence heritage. More broadly, an interpretation of military history that stresses the role of great men, and a very few women, doing great deeds in a great cause, is seen as providing historical legitimation for a prevailing ideology of the national government, which itself wishes to emphasize the importance of the enterprising individual in shaping events. There was certainly a bizarre combination of past and present in 1982, when Portsmouth was the 'flagship' of 'National Maritime Heritage Year' while preparing, waving off and welcoming back the Falklands task force.

Education has always been one of the justifications for the conservation of historic artefacts. It is, therefore, not difficult to admit that heritage has a clear socialization function. However, in practice, the tentative arguments which try to trace a link between defence heritage and a particular dominant ideology are difficult to demonstrate. Individual consumers are likely to experience a wide range of quite different reactions in response to the same heritage, as a result of their different motivations. It is also clear that such heritage can be used in a variety of different ways, often within the same presentation, in support of ideologies other than those discussed so far.

A 'romantic chivalry' approach

This is the most common approach to most medieval military architecture, with attention paid to knights and damsels and to war as a mixture of sport and the social duty of a specific class. It owes much to the nineteenth-century romanticism of writers such as Sir Walter Scott or architects such as Viollet-le-Duc, and is often presented in combination with participation in 'medieval' jousts or banquets.

A cultural separatist or local patriotic approach

In essence, this is a regional or local variant of the nationalist approach, but using the defence heritage to support a separatist identity. The accent, therefore, is strongly upon the role of military architecture and on place associations, in defence against the centralizing power. Urban military architecture is frequently less than ideal for this purpose – more often being representative of the conquests of the centralizing power than the resistance of the locality. Most Welsh and many Scottish castles, and the fortified towns of North Wales, are symbols of military conquest rather than resistance – although examples of the latter can be found, such as the Cathar defences of Languedoc. Towns where military events have occurred can acquire the status of sacred space in a separatist cause, such as the symbolic importance of Guernica to the Basques because of the bombardment of 1937.

A socialist approach

A major contrast with the use of defence heritage to illustrate the noble deeds of great men is an interpretation stressing the everyday lives of the common man, whether a member of the military forces or a citizen suffering the effects of war. The restoration of the fortress town of Louisbourg in Nova Scotia, for example, included not only a reconstruction of the architecture but a repopulation of its inhabitants: visitors are confronted by a 'French garrison', whose state of dress and behaviour (but not smell) is supposed to convey their poor living conditions and demoralization.

Such an approach is not in itself socialist but could be harnessed to ideas of class repression, and its reaction in class solidarity. It might be expected that examples would proliferate in eastern Europe and the Soviet Union, and certainly there is a tendency in those countries to accentuate the defence heritage relating to selected periods in history when rebellion or revolution against the pre-communist established order occurred. However, recent defence history tends to be interpreted in a nationalist rather than international socialist manner, although with a strong accent upon the individual soldier or citizen rather than the influence of great leaders. In practice, some of the clearest examples are found in the heritage presentations of local authorities with left-wing governments in western Europe. In these, castles and town walls are seen as symbols of social and political oppression and used to interpret the situation of the common people who built them, peopled them and lived in their shadow. Norwich castle carries the notable dedication to 'the long struggle of the common people of England for just conditions'.

A technological/aesthetic approach

This is often claimed to be a value neutral approach, in so far as attention is directed to the form of the relict or reconstructed objects themselves and away from the ultimate purposes to which they were put. Defence works become a part of industrial archaeology or architectural history, and are interpreted as a progression of technical solutions to scientific problems, with this striving for functional proficiency leading to perfection in physical form. Fortifications are frequently presented as 'military architecture through the ages', and the weapons of war (from swords to battleships) as studies in metallurgy, ballistics, engineering and the like.

Even the organization and operations of the users of such objects can be approached in such scientific terms, with attention being concentrated on strategy and tactics as an abstract series of geometrical solutions. The purposes of the activities, the causes of the resort to arms, and the effects upon individuals are ignored. Conflict will be described in a neutral terminology which distances these activities from their impacts upon people. War is reduced to a chess game, played for its own sake according to a mutually accepted set of rules, whose outcome is determined by the professional skills of the commanders with little thought for the fate of the individual playing pieces. Although this approach can be found in the interpretation of defence heritage of all historical periods, it lends itself particularly to the period from the Thirty Years War to the Napoleonic Wars in Europe, during which time war was largely seen and taught in the military academies as a 'professional activity' based upon scientific principles, in contrast to the feudal obligation of a particular class as in previous centuries.

Although ostensibly ideologically neutral, this sort of presentation of defence heritage can have ideological consequences. Instruments of war are accorded a moral status equal to similar products of technology and science which were created for peaceful purposes. Thus, for example, HMS *Warrior* is presented at Portsmouth as an example of mid-Victorian engineering achievement, in the same way as similar relics of the period – such as the Clifton suspension bridge or the SS *Great Britain*. This presentation has two possible important effects. First, it encourages a certain loss of sensitivity to individual suffering as a consequence of its casual acceptance implied by such a science and technology of war. Second, it carries the implication that such conflict is not only an inevitable part of human history but has always been a normal activity of a rational profession rather than a bizarre and irrational aberration. These results may have contemporary political consequences through their effect on public attitudes towards defence policies. For example, the possession or use of nuclear weapons can be made more publicly acceptable by reducing them to a set of technical specifications and couching the discussion of their operation in the 'scientific' acronyms and professional phraseology found in Chapter 6.

A peace and international understanding approach

Given the above discussion about the way in which defence heritage is presented in practice, it seems incongruous now to argue that it can be, and occasionally is, used in support of international understanding rather than competition, and for the advancement of peace rather than war. To some, the obsessive interest in the accoutrements of war is clear evidence of an unhealthy trend that can only contribute to a glorification of past conflicts and thereby make future conflicts more likely. As Holmes (1985) has pointed out with the study of military history, which applies with equal force to the incorporation of defence into heritage, an interpretation that stresses the technical side and distances itself from its effects will be inhuman, while one which concentrates upon individual suffering has at best an ultimate numbing effect and at worst encourages an element of voyeuristic sadism.

In support of a pacifist approach it can be weakly argued that, given the existence of this prevailing curiosity about this aspect of our past, some attempt at least should be made to deflect its most undesirable ideological implications towards, if not the horror, at least the futility of war. More robustly, it can be asserted that war will never be prevented by those who know nothing of it and thus it follows that defence heritage has an important educative task in ensuring that the past is not allowed to repeat itself.

Difficult as it may seem, much is being achieved (consciously or not) along these lines. The custom of British Commonwealth forces of burying their dead where they fall has scattered military cemeteries around the world as an impressive part of defence heritage that carries its own message. The currently fashionable thematic presentations at many military sites and museums include 'everyday life' displays which inevitably show the similarity in experience between friend and enemy. German cities face a particular problem in commemorating and interpreting the events of 1933–45. The central areas of many of these cities were destroyed (see Chapter 6), and the very redevelopment is a permanent visible reminder of a past suffering that demands interpretation. Silence frustrates the curiosity of new generations but a nationalist interpretation would be unacceptable. Different cities have adopted different solutions but two examples will suffice. In Kassel, two plaques, side by side on the town hall, relate on the one side the statistics of casualties and property destruction in air raids and, on the other, the local voting figures for the 1933 elections, in which the National Socialist party received the most support. Similarly, in Lübeck, an exhibition displays the results of RAF raids, including photographs of the destruction of the cathedral. Also included, however, is a display of similar photographs of the Luftwaffe raid on Coventry, again including the destruction of a cathedral. In both cases, the events related and objects displayed are no different in themselves from those in hundreds of such commemorations but the arrangement is intended to encourage quite particular conclusions to be drawn about the nature of responsibility for war.

Many of these 'ideologies' will be presented in a partial, mixed and often quite unconscious manner by those claiming only motives of accuracy, authenticity or even entertainment. But that does not make them any the less insistent, nor is a 'neutral' or 'non-ideological' approach possible, being in itself an ideological interpretation of defence heritage.

Dissonant heritage

A special problem arises with the conservation of defence heritage that is strongly and inescapably imbued with an ideology that is currently unacceptable. An extreme case of the dilemma of how to treat artefacts that may be both authentic and historically important but ideologically dissonant, is the contemporary debate in Germany about the architectural relics of the Nazi era, especially the 'Speer' monuments in Berlin and Nuremberg (Selier 1988). Essentially, the same problem exists with the military relics of all previous regimes which are disowned by contemporary state ideologies. Much seems to depend upon the passage of time dulling such sensibilities. The heritage of the Roman occupation of northern Europe creates few problems but the West African slaver castles clearly do. The military relics of a colonial past are interpreted as part of national history in the United States and the countries of Latin America but present problems in more recent colonial successor states.

A partially related set of interpretation problems could be labelled the 'atrocity heritage', of which the clearest examples in Europe are the concentration camps of the 1933–45 period, but which could include a wide variety of relics of the results of violence – from the 'Black Hole of Calcutta'; the civilian massacres of Drogheda, Magdeburg and numerous other cities; to the atomic destruction of Hiroshima and Nagasaki. It is a disturbing fact that atrocity is a particularly popular aspect of military heritage, and its interpretation (and even its preservation at all) poses a number of obvious dilemmas. This heritage can be treated in a purely nationalist way (the sufferings of one national or ethnic group at the hands of another), or in terms of a set of universal moral imperatives (the triumph of good over evil). There is no guarantee, however, that the individual visitor draws the conclusions suggested by the official interpretation and not diametrically opposite ones.

The problems of dissonant heritage, which are especially apparent with defence heritage as a result of its close connection with violence are only a particularly sharply focused part of the wider questions raised in this chapter (and considered at greater length by Tunbridge 1984), of whose heritage should be preserved and interpreted in whose interest.

DEFENCE HERITAGE TOURISM

Symbiosis?

Inherent in the definitions argued above is the idea that all heritage requires a market of consumers, and, in so far as part of this market originates further than a day's journey from the heritage site and travels to it specifically to experience defence heritage, then we arrive at the concept of 'defence heritage tourism'. Viewed from the opposite direction, tourism can be defined as either a set of activities or a package of facilities, undertaken or used by visitors principally for pleasure. In either event, tourism needs 'primary resources' – that is, facilities or attributes of an area that attract visitors to it (Ashworth 1985). Defence heritage can fulfil that need. Thus, redundant, conserved military works and associations become resources within a tourism industry. This might appear to be a logical, obvious and happy symbiosis. Redundant buildings and artefacts acquire a new use – which may imply a new source of financial support – while an important and growing economic activity acquires a new primary resource.

The possibility of such a harmonious, mutually profitable relationship is not questioned here, only its inevitability. Defence heritage tourism is a world-wide success, whether measured in terms of the proportion of redundant defence works which have found a new use from this activity or in the relative size and importance of the tourism industry that has exploited this heritage as a resource. The implications of this relationship, however, are by no means self-evident, nor necessarily harmonious. In order to investigate these implications further, it is thus necessary first to place defence heritage within the wider context of tourism demand and supply, and then examine briefly some of the areas of potential conflict between defence heritage and tourism.

Defence heritage tourism within tourism

The seemingly simple yet fundamental questions about the nature and size of defence heritage tourism cannot be directly answered from the mass of statistics available on tourism. Visitors include international tourists, domestic holiday-makers, day excursionists, business and conference visitors and many more such categories (see, for example, the attempts at definition and measurement in Pearce 1987). The main difficulty from the point of view of the heritage attraction is that users may be drawn from a wide variety of such

categories at any one time and, conversely, any particular visitor will be combining a wide variety of different experiences, whether tourist or non-tourist, heritage or non-heritage, within the trip. Defence heritage tourism therefore is neither a self-contained definable industry nor a particular type of holiday motive that can be isolated and measured in simple terms.

Estimates can be made of the importance of urban destinations within the global totals. Cities, most especially the major world cities, are the most important type of destination for foreign tourists, and generally the second or third most important destination for domestic holidays in most countries (Ashworth 1989). To that could be added the importance of cities as destinations for day excursions, by both residents of the region and by those holidaying in other sorts of environment within the region. A major motive for this attraction to the cities is heritage, – a theme pursued in detail in Ashworth and Tunbridge (1990). Within heritage, defence heritage plays a major role. These assertions are supported more by conventional wisdom than statistical demonstration, and empirical evidence is episodic. In the handful of national-scale studies of the motives of incoming foreign tourists, 'history' or 'heritage' is stated as the most important in Britain (English Tourist Board 1981) and France (Garay 1980). In studies of the promoted images of countries as tourist destinations, it is again 'history' that is generally the most important element in such images (Dilley 1986).

A step further is to consider not merely the overall size of the heritage tourism market but its nature – seeking answers to the question, 'What sort of person is the heritage tourist?' Again, the evidence is piecemeal but the visitor to heritage sites can be generally described as being middle-aged, child-free and of above average income and education (NRIT 1988). Although such a profile may account for the numerically, and probably also financially, most important segment of the market, there are many other identifiable categories – ranging from school parties, unattached young people with 'wanderlust', organized tours of pensioners and many more 'speciality' groups, including in this context those of military veterans, and amateur-history and battlefield 'buffs'.

Thus we arrive at the very general conclusions that the tourism market from which defence heritage draws its customers is large, growing, extremely varied in terms of motives and visitor characteristics, and composed of distinct segments.

One further set of comments can be made about visitor behaviour which is important to the operation of the heritage sites themselves. Tourists attracted by the resources of the cities and those motivated principally by heritage have a distinctly higher pattern of daily spending than most other groups, are more likely to choose serviced accommodation forms, and are less prone to seasonal variations in the timing of visits. Equally, however, the total length of the trip tends to be shorter than the average for all holidays and the length of stay in any one place is particularly short. The average length of stay in an individual city, even one of the large world cities, is around 2–3 days (Burtenshaw *et al.* 1981), and in any one heritage site or museum the same number of hours.

When viewed from the supply side, much the same difficulties of estimation and general conclusions emerge. There are samples of statistics collected at individual sites that can be used to indicate the importance of defence heritage sites within tourism as a whole. For example, the *English Heritage Monitor* produces annual counts at major individual tourist attractions, which, while neither comprehensive nor necessarily counting only tourists, narrowly defined, nevertheless gives some indication of importance. Among the 24 'top' historic properties (i.e. those with more than 200,000 paid admissions in 1987) there are 6 castles (the 'Tower', England's most popular historic attraction, and the castles of Windsor, Warwick, Leeds, Hever and Dover), 3 warships (*Mary Rose*, *Victory* and *Belfast*), and 2 war commemorations (Marlborough's Blenheim, and Churchill's Cabinet War Room) – see Department of the Environment 1988.

However, such statistical information is partial, in so far as defence heritage is only one part of a wider heritage experience, and rarely distinguished from it by most visitors who move from castle to town hall, or between defence and non-defence exhibits in a single museum. Similarly, if defence heritage is to be regarded as a tourist resource, it must be seen as only one part of a much wider total package of resources that are combined to form the tourism experience. The important implication of this, together with the typically short length of stay mentioned earlier, is that heritage tourist attractions must be considered in terms of networks rather than as individual sites. Very few sites will be attractive enough to alone provide the motivation for the visit or to hold the attention of the visitor for more than a few hours. To be successful tourist attractions, defence heritage sites must be located within appropriate networks of both other tourist attractions and secondary tourist support facilities. The composition and nature of such networks are, therefore, critical. Equally, however, they can be extremely varied in terms of spatial scales and the characteristics of the composing elements.

A few examples of different sorts of such networks at different spatial scales may illustrate some of the general characteristics. On the urban scale, in Boston (see Figure 8.1) the visitor's 'landscape' of the historic city is composed of a town trail (the 'freedom trail') along attractions, many of which are defence related – a battlefield (Bunker Hill), a naval dockyard, two warships (*Constitution* and *Cassin Young*) and the site of a riot (Old State House), linked by short ridges of interest (the corridors of movement, lined with various catering and tourist shopping facilities) and surrounded by a tourist *terra incognita* of the rest of the city. The location of defence heritage sites in relation to the tourist circuits in this historic city is an obvious major factor in their success in attracting tourists.

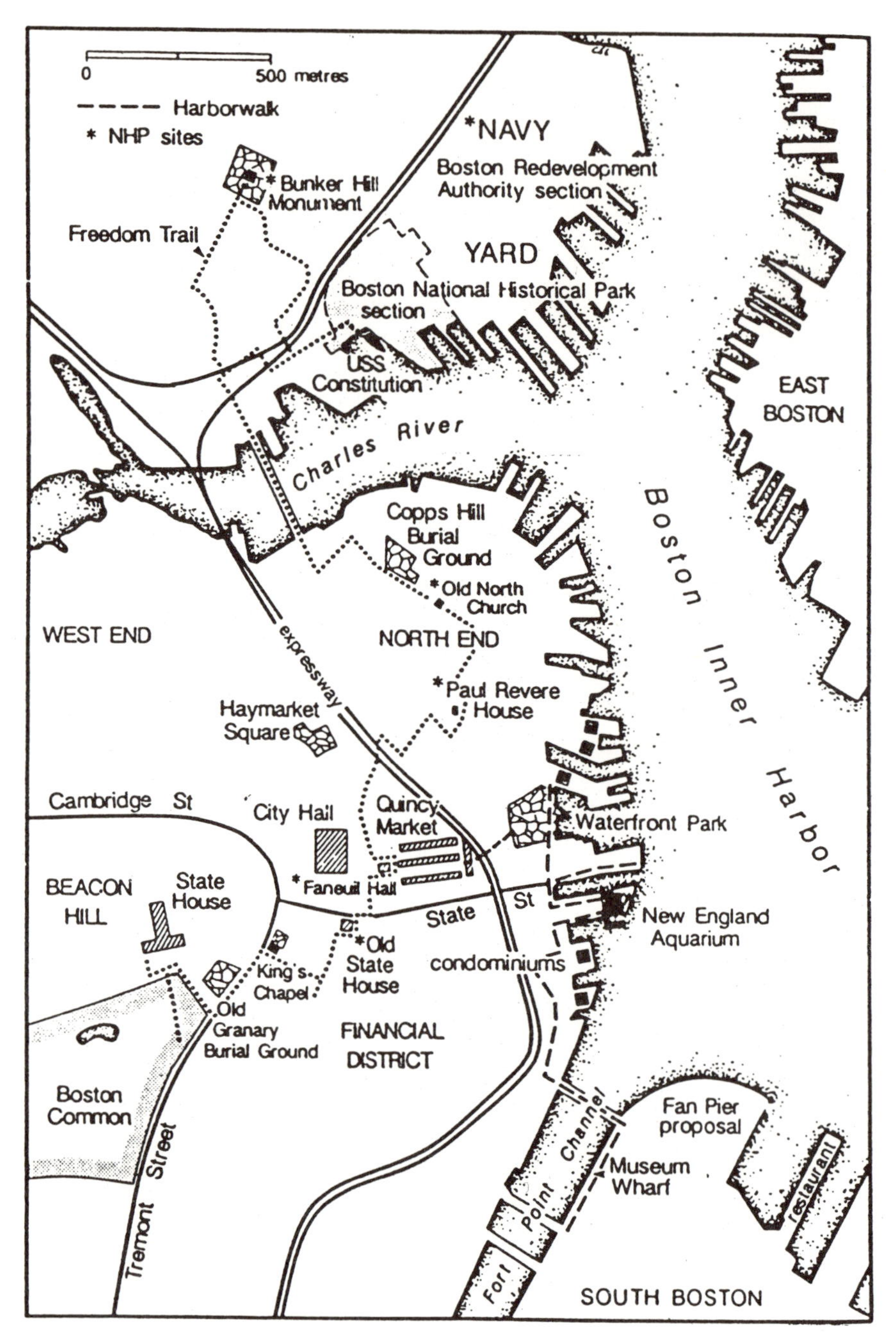

Figure 8.1 Defence heritage in Boston, Massachusetts

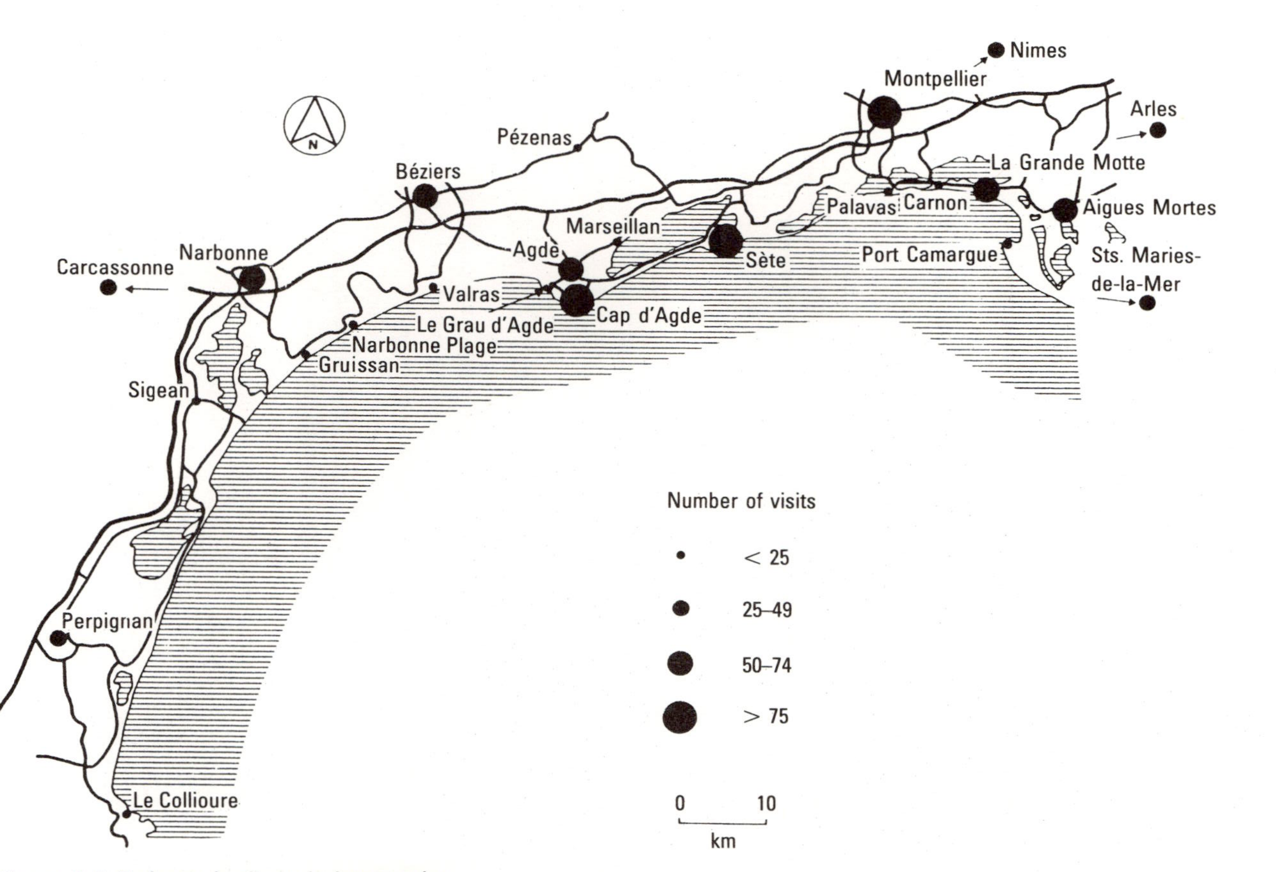

Figure 8.2 Defence heritage in Languedoc

On a regional scale, Figure 8.2 shows a seaside resort region, within which historic cities are either serving as small specialized excursion destinations by offering mainly defence heritage (such as the walled 'crusader' town of Aigues Mortes and the late-medieval fantasy fortress of Carcassonne) or are providing generalized urban facilities, including heritage attractions (such as Narbonne, Sete or Montpellier). Defence heritage is one part of a wider use of history which, in turn, is associated functionally and spatially within a holiday region dominated by other sorts of tourism attraction.

On a still wider scale, a single urban cluster of heritage sites can form part of a regional, national or international circuit of tourism attractions. The sorts of itineraries to be found in many package tours could include: the historic cities of Flanders; Scotland's heritage (with Edinburgh castle, some Highland forts, a few battlefields from the 1745 rebellion, and the massacre at Glencoe); or even 'capitals of Europe'. The important point is that tourism 'consumes' heritage sites and centres extremely rapidly, as parts of much broader holiday packages. In addition, defence heritage, although forming an important ingredient, is rarely presented distinct from heritage as a whole. Furthermore, specialized defence heritage holidays (e.g. 'battlefield tours' or 'castles of the Rhine'), although increasingly popular, are still exceptional.

CONFLICTS BETWEEN DEFENCE HERITAGE AND TOURISM

One of the most obvious and well publicized of such conflicts results from the simple spatial coexistence of historic monuments and the large number of tourists enjoying them. Physical damage, whether intentional or not, is perhaps less significant with many items of defence heritage which are intrinsically robust. However, the physical presence of large numbers of tourists (who are by definition in holiday mood), together with their ancillary needs for refreshment, parking and souvenirs, can have a disastrous impact upon any ambiance that the monument or site is intended to convey. This is particularly so with defence heritage, in which the inevitably sombre and serious tone of battle is reduced to the atmosphere of a carnival. It is difficult to find much of MacCannell's (1976) 'sacred space' at major defence heritage tourism sites. Tourists may be strongly attracted to the conserved relics of the past but are themselves firmly a part of the present; as such, they require modern support facilities which only rarely can be successfully incorporated into the historic attraction. The amelioration of these sorts of conflict is essentially the task of planning and management. Skilful practice can raise the carrying capacity of sites but the fundamental point here is that tourism, while providing a use and justification for historic land uses in cities, simultaneously and inescapably makes other and less welcome land-use demands.

A related but less visually obvious conflict occurs as a result of a fundamental discrepancy in the economic and organizational relationship between the visitor and the heritage site. The defence heritage, alone or in combination with other attractions, may have provided the primary motive for the visit but it is highly unlikely to be either financially self-supporting or to have been created principally for this purpose. Many defence-related places and objects in the city can be enjoyed either free of charge (such as city walls, the exteriors of buildings, battle-sites, etc.) or for a below-cost fee (such as most museums and publicly owned castles). They are by design, or by necessity, public goods – generally provided by public authorities for communal goals. The visitor pays instead for the costs of secondary support services, usually provided by private commercial organizations. Thus the commercial tourism industry is dependent upon a heritage resource that it has not created, does not manage, and only indirectly finances through taxation. Conversely, defence heritage has been created for a particular set of motives, and is managed and financed by organizations that have little to do with its important tourism market. This paradox may not be evident in purpose-built and managed heritage tourism sites but these are rare in the context of the multifunctional city.

The third important general area of potential conflict stems from the selectivity of the tourist, which results in tourism making an intensive use of only a small fraction of the total conserved resources. In particular, it is likely that tourism will only be interested in the larger, spectacular 'marked' buildings and sites. This selectivity is only a spatial aspect of a wider question of the relevance of heritage. Tourism's answer to the question, 'whose heritage?' is, unambiguously, 'the heritage relevant to the visitor', and this has a number of important implications.

The visitor may have a limited knowledge of both defence history in general and the individual heritage site in particular. In order to match expectations, much of the complexity and richness of the past must be selectively simplified – in practice, 'bowdlerized' into what Ford (1978: 22) has called a 'sanitized, idealized past'. In addition, the visitor may have not only a simpler version of history but a different historical experience. This implies selection of sites, buildings and interpretations, orientated towards the historical experiences and expectations of the visitor, and, conversely, an underemphasis on other aspects of heritage. The problem of tailoring the heritage to the nature of the market is not that there is a conflict with authenticity, rather that the heritages of visitor and resident may be sufficiently different as to create conflicts about the purpose and practice of conservation.

In particular, this is likely to be the case where defence heritage interpretation has been based upon nationalist or regionalist ideologies, and where the tourists originate from outside the country or region. The 'whose heritage?' discussion can be at its most controversial when this question arises in sensitive post-colonial societies – such as in the West Indies, where colonial forts and defences can be exploited as major tourist attractions relevant to the historical experience of the potential visitors but which commemorate periods of history which often do not fit comfortably into local prevailing ideologies.

FROM RESOURCE-BASED CONSERVATION TO DEMAND-BASED TOURISM

As argued earlier, preservation and reconstruction are opposite ends of a seamless continuum, rather than alternatives. Preservation that has elected to find a justification in heritage is committed to a market which, in turn, needs management and interpretation. It is an almost imperceptible step from the selection and presentation of what is preserved to the reconstruction of what once existed but by chance has not survived. One further step is the construction of what might, ought, or could have existed but actually did not; and, one step further, the construction of what the visitors expect to have existed but actually could not. The difference between the extremes of an archaeological site and a Disneyland fantasy is obvious, but the intervening stages are not.

The town walls of Norwich (and, more so, York) have been repaired and largely raised to their former height, and so much woodwork has been replaced on HMS *Victory* at Portsmouth that only a small proportion of the ship is original. In both cases, the 'authenticity' of the heritage experience to the visitor – in the sense of conveying 'how it was' – is actually increased by reconstruction. The fortress town of Bourtange, on the Dutch-German border, has gone one stage further – by re-creating the ramparts and ditches that had been dismantled and filled in 100 years earlier. The fortress town of Louisbourg (Nova Scotia), meticulously destroyed in 1759, has been equally painstakingly rebuilt, at a cost many times more than the original. The physical relics are now completely artificial but the heritage experience remains 'authentic'. On the battlefield of Waterloo, the visitor can experience the site – the authentic spot where this event occurred – or a choice of audio-visual presentations. There is little doubt that the latter is more 'real' to most visitors, in the sense of conveying a more vivid impression of what happened in 1815. The distinction between these examples (which remain resource-based in terms of their location and their 'historic product') and demand-based sites (whose existence is a response to visitor demands rather than the characteristics of the resource) can be small. Many of the conflicts (discussed earlier) between conservation and tourism, arising from the pressures of visitors, are ameliorated by exploiting this difference in 'reality' between the two. The 'real' Hadrian's Wall of the archaeologist is protected by the 'real' tourism experience of the reconstructed Vindolanda Fort.

There can be a wide discrepancy in amount, type and location of the supply of heritage artefacts and the demand to experience them. One solution is the construction of a heritage site to satisfy the demands of visitors at a location that accords with that demand, rather than any relict resource allowing the heritage experienced to be free from the constraints of chance survival in 'wrong' locations. For instance, Den Gamle By, at Aarhus in Denmark, and the historic town (*cita historico*) of Turin, were built as examples of medieval town quarters in order to present the medieval urban scene as it might, or ought to, have been – if such an architectural relic had survived. Relatively newly settled countries have an obvious problem in this respect, and thus tend to define heritage in terms of the experience which may be induced by facsimile or theatre rather than the conserved object.

Finally, the assumption that the supply of conserved artefacts is immobile can be relaxed for some categories of military relic. Apart from the obvious demand orientation of museums, whose exhibits are assembled and transported to the customer rather than the reverse, defence heritage is notable for the importance of relatively large equipment and vehicles that are intrinsically mobile. The most spectacular example of this is the warship, whose preservation and use as the centrepiece of a military heritage presentation could conceivably be located anywhere, or, for the larger examples, anywhere with access to navigable water. Portsmouth's *Mary Rose*, *Victory* and *Warrior*, London's *Belfast*, Yokahama's *Mikasa*, Stockholm's *Wasa*, Leningrad's *Aurora*, Roskilde's Viking longships, and many more, are all major tourist attractions which could have been located elsewhere. Facsimile replicas, such as Plymouth's *Golden Hind* or Piraeus's trireme, are almost equally effective and intrinsically demand-orientated.

In North America, the late eighteenth-century frigate *Constitution*, together with the Second World War destroyer *Cassin Young*, forms the centrepiece of the Charlestown Navy Yard, which became a part of the Boston Harbor National Historic Park in 1974. This national heritage site is a complex of conserved and reconstructed waterfront buildings, related exhibitions and displays, and visitor facilities. It has been influential in attracting new commercial and residential land uses to the surrounding parts of the inner city, and thus has become a renowned textbook example for the revitalizing of what was previously a discarded waterfront zone (Tunbridge 1987). There are many similar examples of the use of historic ships as centrepieces for the conversion of semi-derelict harbour areas to heritage centres as part of revitalization projects: the *Sackville* at Halifax's 'Privateers' Wharf'; the *Olympia* at Philadelphia; the *North Carolina* at Wilmington; and the *Massachusetts* at Fall River.

The important point in this context is that this account of the re-use of redundant defence works in cities has now arrived at a quite different set of assumptions and values than were present at the beginning. What began as the search for new functions for existing structures has now become the use of those structures as vital instruments in urban economic revitalization. The urban economic well-being that at one time insisted upon rapid demolition and re-use of the resulting space can now be equally strongly committed to not only the conservation of defence artefacts but their reconstruction. Redundant defence works – whether as obstacles to development, as the past to be preserved, or as an exploitable resource of the heritage industry – are an integral and important part of the form and functioning of cities.

Chapter 9

The city and defence

AN UNEVEN TREATMENT

The intention of the book, as stated in the introduction, was simply to trace the existence of links between collective, organized defence as an activity and aspects of the form and function of the contemporary city. If such links could be discovered, and were traceable, demonstrably important in comparison with other variables, and explainable in a coherent manner, then the ideal result would therefore be the emergence of both a 'geography of urban defence' and a 'defence geography of the city'. The former should provide a framework for understanding the role of cities as places within defence studies, while the latter would be an 'aspect' urban geography, alongside many similar studies of the contribution of specific sets of variables to the shaping and spatial patterning of urban settlements.

A concluding chapter should assess to what extent these intentions have been realized in the preceding chapters. In so far as the coverage of the content falls short of the norms of comprehensiveness and balance – in terms of either aspects of the defence activity or types and locations of urban environments, or in so far as the structuring frameworks of analysis are insuffіcently internally logical or effective as methods of explanation – then these departures from the ideal however inevitable, need identifying, explaining and evaluating.

Two main difficulties have been encountered. First, the field is extremely broad: there are simply a very large number of quite different ways in which defence and the city are related. Second, the relationship between defence and the city is inextricably related to many other links between activities and places. They are, therefore, not only particularly difficult to isolate but may, in practice, only be understandable in a much wider explanatory context that strays far beyond both the defence activity and the urban scale.

A consequence of the first difficulty is an uneven coverage: some of the relationships have been treated at length, while others have received more cursory attention. Studies of the effects on cities of nuclear attack and, for instance, the resulting defensive strategies and provision, have been adequately examined, from various viewpoints, elsewhere, and have only been reiterated here as part of the longer experience of the 'hostage city'. Similarly, there is an extensive literature on the economic impacts of defence installations and equipment procurement, which, although of considerable importance to the economies of particular cities, is only satisfactorily explainable in terms of national economic and defence policies, a discussion of which would go far beyond the scope of this book. Frequently, an unevenness in coverage reflects a similar unevenness in the state of existing knowledge and its dissemination. Quite simply, very little is known, or at least is published, about the city as battlefield terrain. Consequently, much of the analysis is drawn from widely diverse accounts of conflict and the conclusions are necessarily speculative. In contrast, there is a growing interest among academics and city managers in the problems of internal defence against insurgency, public lawlessness, and the threat of violence. This has generated a substantial body of analytical work, which requires only to be reduced to a manageable form and to relate both to wider defence issues and the characteristics of cities. Similarly, the history of urban fortification rests upon an enormous quantity of descriptive historical chronology and case-study material. The difficulty with the ordering of such material lies in relating it to the contemporary city, through either the impacts of past defence works upon present morphological patterning or through the 'commodification' of the past and the creation of heritage as an urban economic resource.

The second general difficulty is that of sufficiently isolating the defence activity from other functions for investigation, and thereby shaping a distinctive 'defence geography of the city'. This problem is so pervasive in each of the above chapters as to need more detailed consideration below. However, the overall conclusion that can be drawn here is that the nature of the defence function renders it peculiarly difficult to separate from the wider characteristics of the place which is both being defended, and engaging in the defence. Frequently in the above chapters particular elements in the defence of the city were more easily explained in the context of quite different aspects of the urban geography than by reference to other aspects of defence. In other words, although defence can be shown to be an important variable effecting the evolution of urban form and function, and thus plays an essential and generally neglected role in urban geography as a whole, it is intrinsically difficult to group these diverse impacts into a coherent urban defence geography.

It also became clear in a number of chapters that the view of the relationship between defence and cities from urban studies is quite different from that taken by military science. Consequently, it has not been difficult to demonstrate that defence is an important variable in many aspects of urban development in many types of city while conversely it has proved much more difficult to argue that the role the city has played in defence studies has been more than a marginal one in exceptional circumstances. This imbalance may reflect a fortuitous disciplinary bias or a fundamental reality in the relationship but, in either event, the intention to use this topic as a bridging relationship has resulted in a discernible asymmetry in treatment, which is traceable to these quite different assessments.

DEFENCE IN URBAN GEOGRAPHY

It is insufficient just to point out, through selected examples, that there are important links between urban defence and other aspects of the city. The argument running through the book is that such links are not only of general rather than particular importance but inextricable from their wider context. The most pervasive of these contexts can be outlined briefly.

Urban form

It is the effects of defence upon urban form and morphological structure, as well as upon the site aspects of urban location, that have been the most persistant over time and, being frequently clearly visible in a dramatic way, the most obvious. This very visibility, and the resulting relatively lavish attention paid to these features by commentators, has dangers in itself. The most unfortunate is that this aspect of the relationship has tended to overshadow other less visible but equally important functional links. A review of the existing literature could easily conclude that the impact upon urban form was the only important effect of defence upon towns, and occurred only in a few spectacular examples. In turn, this not only encourages a neglect of other aspects and applications to cities as a whole but encourages such studies to be most logically approached historically. This, in turn, suggests that defence is a function of the past, whose significance has diminished in the cities of the modern world. I have no doubt that the title of this book evokes images for many of castles and city walls rather than anti-ballistic missile (ABM) silos and anti-terrorist surveillance systems.

More subtly, in a number of chapters it has emerged that the effects of defence on morphology could be related to more than the physical impacts of fortifications, extensive though these could be. The influence on the evolution of urban form of the provision of defence services, garrisons, supply facilities, and the like, could be as important, if less visually dramatic. Similarly, the threat of insurgency in its various guises is a variable in urban design whose significance has not diminished over time. Clearly, the creation and maintenance of urban defence works has had an enormous influence upon urban structure, but their demolition or re-use has been equally significant. The dismantling of fortification systems raised issues of urban planning and management on an unprecedented scale in many cities, and the response to this challenge has had an importance to the development of town planning that is only slowly being appreciated. Similarly, the transformation of the past into heritage has led to a re-evaluation of historic defence works and associations, so as to complete the circular relationship between defence function, urban form, and defence heritage function – whose implications are still to be understood.

Urban economies

Even in the chapter considering the fortification of cities, a recurrent theme was that these should not be seen only as architectural objects and physical structuring elements in the city: they were also an integral part of the economic functioning of cities, and a resultant of various economic costs and benefits. Most obviously, defence works could be shown to be a factor in the spatial zoning of economic activities. More fundamentally, the questions of whether, and to what extent, to defend a city, and, using whose resources in that defence, with which results to whom, all ultimately received economic answers.

It is easy to demonstrate the economic significance of defence – whether through equipment-procurement, the siting of garrisons, bases or other defence-related services – to the special case of the defence town (the martial metropolis), since this is usually characterized as an exceptional town, with a distinctive geographical character. Defence, however, has had a far wider, if more diffused, impact upon urban economies in general, whether as part of national, externally orientated military systems (in investment in security from aerial attack or from internal insurgency) or as an economic resource.

Urban society and culture

The defence function can be shown to be a significant variable influencing the detail of the social composition and spatial social structure of towns. Fortifications were not only an influence upon the evolution of the physical morphology but also a long-term and widespread factor in the development of the social and occupational patterning within the city. Military personnel, by the nature of their profession, are frequently aberrant in their demographic and social characteristics, attitudes and patterns of behaviour, possessing an internal social cohesion (which can amount to a military caste) that necessarily isolates them from the wider society. Those particular towns dominated by defence personnel, their families, pensioners and those economically dependent on servicing them, are likely to reflect these characteristics and thus be distinctive within the national society.

A more complex pattern of cause and effect was evident in the studies of defence against internal insurrection – where there was a constant triangular relationship between insurgency, the physical morphology and the social geography of the city. In many of the cases examined, the authorities assumed the existence of a direct link between urban areas with distinctive ethnic or economic characteristics and a tendency to insurrection – whether public lawlessness, riot or guerrilla operations. Changes in the physical morphology were then used as an instrument for discouraging such insurgency or rendering it less effective. There are only a few examples of the more difficult policy of deliberately attempting to alter the social ecology of areas by removing or implanting social groups to favour either the security forces or insurgents. Both insurgency and counter-insurgency have tended to use the city's physical patterning tactically and its social patterning strategically – the latter only on occasion, and that frequently by chance.

All these are particular traceable relationships between the social geography of towns and their defence. There is, however, a far more fundamental set of questions, the answers to which can rarely be demonstrated clearly and which go well beyond the scope of this book. The term 'the city' has been freely used in the above chapters as if it was a homogeneous and coherent entity defending itself or being defended, whereas it is a set of heterogeneous social groups cohabiting in a particular physical space. Frequently, the questions, 'who was defending or attacking the city?', 'in whose interest?', and 'at whose cost?', could be answered in terms of particular groups within urban society. Even the abandonment of defence works and their translation into heritage raised the important issue of 'whose defence heritage, interpreted and presented in whose interest, propagating what vision of society?' These are important, pervasive and omnipresent issues and the fact that they can only be touched upon briefly above is explainable by their central significance. This makes defence only one aspect of urban society, and the pursuit of all these issues would lead once again to studies of the city as a whole beyond the scope of this volume.

Urban politics and power

Similarly, the relationship of defence to the geography of urban politics and power raises much wider issues, but in this case the considerations of defence lead rapidly and inevitably beyond the confines of the individual city. In many of the defence situations considered above, the city was in effect a political hostage to wider circumstances – with a severely constrained capacity to react in the protection or furtherance of local interests. Decisions made on a quite different spatial scale, for wider economic or strategic reasons, determined the security, prosperity, or even survival of the city. The extreme case of annihilation, or threat of it, being used as an instrument of national strategy recurs in Melos (416 BC), Haarlem (1573), Magdeburg (1631), Rotterdam (1940) and Hiroshima (1945), and merely became both more general and technically easier to undertake in the nuclear age. Equally widespread, if less comprehensive, is the terrorist use of the city as symbolic example and means of publicity for political causes, that may not only be far wider than the urban victim concerned but may actually have no obvious relevance to it.

URBAN GEOGRAPHY IN MILITARY SCIENCE

It has generally proved far easier to demonstrate that defence has played an important role in many aspects of the city than to show that the city has played a central role in military science. This may well be true more generally – in that, although there is a 'geography of defence' it is likely to be of only marginal significance to the conduct of military activities compared with other variables, and is frequently reduced to either an inventory and classification of the defence characteristics of terrain at the tactical level or a spatial geometry applied to movement at the strategic level. If this is so, it would help to explain the relative absence of such 'geographies' in defence literature, and, in particular, the difficulty of shaping a specifically urban geography of defence. The links that have been drawn in the above chapters have tended to operate through three aspects of defence – namely, weapons technology, military organization and defence policy.

An analysis of the interrelationships between these three factors is central to almost any study of defence. Its comprehensive pursuit would therefore lead well beyond the more limited scope of the arguments advanced above. It is necessary, however, to stress two points. First, the focus of this book has concentrated upon isolating and outlining those intrinsically urban characteristics of cities that have a bearing on the conduct of military operations. (Whether such characteristics are to be regarded as significant or not can only be determined by a consideration of this wider context). Second, if, and when, these urban variables do play a part in defence, then the effectiveness of such a role and how in detail such variables are used, can only be assessed in relation to these other aspects of military science. These interactions are likely to be multi-faceted and complex. Both these points are not just obvious disclaimers that much of importance to military science has been left uninvestigated or merely touched upon in passing, they go a long way towards explaining the asymmetry in the link between defence and the city that was asserted above.

The relationship between the development of weapons technology and cities was only exceptionally explainable in terms of a simple technological cause and identifiable urban effect. More usually, there were triangular and reversible links between: changes in weaponry; the existence of the need, capacity, and creativity among military organizations to use such innovations; and the form and functions of cities. Defence was rarely technology-driven, and cities were generally more than passive reactors to changing weaponry. Occasionally, particular scientific breakthroughs (whether the invention of the sling-shot or nuclear fission) have had an obvious, intentional and immediate set of effects upon cities. More usually, specific perceived military needs arising from particular defence circumstances called forth an ingenuity in adapting a well-tried technique or applying one developed for quite different purposes. Neither barbed wire nor the airplane had any defence function until military necessity recruited them. The long-drawn-out interaction between the design of urban fortifications and the use of gunpowder artillery illustrates the complexity of the relationship. The initial adoption of gunpowder was in field rather than urban-siege operations, and for some time the latter proved more effective in the defence of cities than in attacks on them. Artillery did eventually have a profound effect upon the design of urban defence works, but over a period of some 400 years of evolution, within which a wide range of other developments in chemistry, metallurgy, ballistics, and electronics were applied to military problems of communications, transport, combined-arms organization and skills training.

Although the evolution of weapons technology, and the engineering and architectural reponses to this, have dominated not only urban fortification but many of the other themes related above (such as aerial bombardment and city annihilation), these should not be considered in isolation from the organization of defence forces. The outcome of military operations was dependent not only upon the weapons and works available but upon the capacities of people to use them effectively, and to direct and manage this use. The extent to which cities fell to assault or resisted it was as much dependent upon organizational factors as upon architecture. Indeed, advances in design and engineering generally outran advances in organization, leadership and training. 'Impregnable' defences (whether eighteenth-century Louisbourg or twentieth-century 'Vesting Holland') fell with nonchalant ease. In the cities of Europe the aerial bomber did not fulfil the expectations that had been placed on it. The lavish supplying of internal security forces with advanced electronic communications and surveillance equipment and a battery of anti-personnel weapons was rarely a critical factor in their success or failure in urban anti-insurgency operations. It is not surprising, therefore, that practical experience more often than not provided lessons in the value of such human factors as leadership or morale, rather than a reliance on military technology.

How cities were used in military operations was dependent not only upon the intrinsic characteristics of the cities themselves but also upon the nature of the chosen defence strategy and policy. In summary, cities offer the exchange of good defensive terrain for poor command control: short effective weapon ranges for poor mobility. Protagonists of a strategy of mobile warfare, whether Napoleonic or Blitzkrieg, will have little use for cities as battlefields; positional strategists of seventeenth-century siege warfare will. The chosen strategy determined the role of cities more often than the reverse. This in itself is not as surprising as the failure of the growing urbanization of the world to be reflected in a parallel growth in the importance of the urban factor in military science. In every century except this one, urban locations have been avoidable battlegrounds except in a few exceptional regions. The modern concentration of people, production and power in cities renders this no longer the case, and the failure of contemporary military science to come to terms with cities is thus harder to explain.

THE PAST, THE PRESENT AND THE FUTURE

The proposition that we live in an unsafe world (in which the security of person and property is under an ever-present threat of violent assault) would have been self-evident to every generation except the current one born in the western world since 1945. It is easy to appreciate that defence was a continuous and primary preoccupation of the city of the past and that the relict effects of past priorities have left an impact upon the modern city. However, the need for defence in the city of the present – and, more so, the future – cannot be so easily assumed. In other words, questions must be raised as to whether the subject of this book is of increasing or diminishing importance. Is the city becoming less or more safe for its citizens? Is defence against deliberate threats to life and property of growing or diminishing concern? The answers to these questions (at least as they have emerged in the arguments of this book) are partial and, on occasion, contradictory.

In terms of documented and reported violence, the city in both the western and less developed worlds appears to be increasingly the stage for demonstration, riot, repression by security forces, terrorist outrage, and public anti-social behaviour. Conflict stemming from differences of political philosophy, regional and national identity, economic wealth, or just individual frustration are acted out, and widely publicized, in the city. Whether the amount and intensity of such violence is more or less than in other historical periods matters little in comparison with modern awareness and concern, which results in increasing collective and individual responses to this image of the city, especially its inner areas.

On the international scale there is uncertainty about the predictive value of the last 45 years. This period of world peace, or, more accurately (considering the numerous post-colonial or great-power, surrogate 'savage wars of peace'), of the absence of major, world-scale conflict, can be regarded as a brief, curious and unique interlude in world history – explainable by a combination of the special political and technological circumstances of the mid-twentieth century. The two superpower, nuclear stand-offs produced a peace, which, although armed and precarious, was a peace which neither side had an interest in disturbing. The late 1980s proved to be a period of rapid, unexpected and unclear political (and thus military) realignment, the ramifications of which are impossible to trace. In China, a degree of economic liberalization was followed by internal political repression. In the Soviet Union, there are clear signs of both economic and political change, which have had repercussions on the deployment of

Warsaw Pact forces and on the role and future of not only the Warsaw Pact but also, in reaction, of NATO. The military scenarios which have dominated defence thinking and preparedness for war, especially in Europe since 1945, will be less relevant in the future. The unification of the German Federal and Democratic Republics and the political restructuring of eastern Europe provide quite different scenarios. Topics such as the future political role of Germany, the nationality question in the Soviet Union and some east European countries, the rebirth of 'Mitteleuropa' and the development of separate foreign and defence policies in the states of eastern Europe, the 'decoupling' of the defence commitments of the United States and Canada from western Europe, are all back on the agenda of international relations after a lengthy absence. The possibilities and pace of such changes and the defence implications of all or any of these items is quite unpredictable.

It could be that the replacement to the two-power world would be less intrinsically secure. Shifting balances within a multi-power international system, ethnic and regional rivalries within and across national boundaries, the persistence of wide differences in economic welfare could all be central features of near-future scenarios – in which case the various defence roles of the city outlined above will receive a set of new examples that could soon be added to those in this book. Alternatively, it could be that the current trends can be realistically viewed as a prelude to a new world order. In this, war would have no place in the settlement of international differences, economic resources would be diverted from defence to welfare, and the political aspirations resulting from cultural differences and economic disparities would be resolved without recourse to violence. If this were the case, cities would, for the first time in the history of their existence, lose their defence functions, and the bulk of this book could then be confined to historical geography and heritage tourism.

References

Abbott, C. (1984) 'Planning for the home front in Seattle', in R.W. Lotchin (ed.) *The Martial Metropolis: US Cities in Peace and War*, New York: Praeger.

Allen, K. (1976) *Big Guns*, Hove: Firefly Books.

Anderson, W. (1984) *Castles of Europe*, Ware: Omega.

Anon. (1982) *De Geschiedenis van de Vesting Bourtange*, Hoogezand, Netherlands: Stubeg.

Argan, G.C. (1969) *The Renaissance City* New York: Braziller.

Ashworth, G.J. (1985) 'The evaluation of urban tourism resources', in G.J. Ashworth and B. Goodall (eds) *The Impact of Tourist Development on Disadvantaged Regions*, Socio-Geografisch Reeks 35, GIRUG, Groningen 37–44.

—— (1988) 'Marketing the historic city for tourism', in B. Goodall and G.J. Ashworth (eds) *Marketing in the Tourism Industry*, London: Croom Helm.

—— (1989) 'Urban tourism: an imbalance in attention', in C.P. Cooper (ed.) *Progress in Tourism, Recreation and Hospitality Management*, vol. 1, London: Belhaven.

—— and Schuurmans, F. (1982) 'Patterns of commercial activity in the central area of Perpignan', *Serie Veldstudies* 4, Groningen: GIRUG.

—— and Tunbridge, J.E. (1990) *The Tourist-Historic City*, London: Belhaven.

—— White, P. and Winchester, H. (1987) 'The red light district in the west European city: a neglected aspect of the urban landscape', *Geoforum* 19(2): 201–12.

Ashworth, L.M.X. (1990) 'The 1945–9 Dutch–Indonesian conflict: lessons and perspectives in the study of insurgency', *Conflict Quarterly*, 10(1): 34–45.

Aston, M. and Bond, J. (1976) *The Landscape of Towns*, London: J.M. Dent.

Aurousseau, M. (1924) 'Recent contributions to urban geography', *Geographical Review*, 14: 444–56.

Bak, L. (1987) *Het vuur van de vergelding: de geschiedenis van de luchtoorlog boven West Europa 1939–45*, Baarn: Hollandia.

Baldwin, S. (1932) Speech to the House of Commons.

Banks, A. (1973) *A World Atlas of Military History*, vol. I, London: Seeley.

Bastier, J. (1984) *Geographie du Grande Paris*, Paris: Masson.

Bateman, M and Riley, R.C. (eds) (1987) *Geography of Defence*, London: Croom Helm.

Beauregard, L. (1972) *Montreal Field Guide*, Montreal: Le Presse de l'Université de Montreal.

Belgian Ministry of Foreign Affairs (1941) *Belgium: The Official Account of What Happened 1939–40*, London: Evans.

Bish, R.L. and Nourse, H.O. (1975) *Urban Economics and Policy Analysis*, New York: McGraw-Hill.

Blair, P.H. (1962) *An Introduction to Anglo-Saxon England*, Cambridge: Cambridge University Press.
Boiten, L. (1988) *Groningen op Weg naar de Moderne tijd*, Groningen: Walters Noordhoff.
Bradbeer, J.B. and Moon, G. (1987) 'Defence town in crisis: the paradox of the tourism strategy', in M. Bateman and R.C. Riley (eds) *Geography of Defence*, London: Croom Helm.
Bradford, E. (1985) *Siege – Malta 1940–3*, Harmondsworth: Penguin Books.
Brand, A.J. (ed.) (1989) *Oorlog in de Middeleeuwen*, Hilversum: Verloren.
Brand, H. and Brand, J. (1986) *De Hollandse Waterlinie*, Utrecht: Veen.
Brice, M.H.C. (1984) *Stronghold: a History of Military Architecture*, London: Batsford Books.
Brodie, B. (1959) *Strategy in the Missile Age*, Princeton N.J.: Princeton University Press.
Brown, S. (1985) 'Central Belfast's security segment: an urban phenomenon', *Area* 17(1): 3–9.
Buchan, J. (1934) *Cromwell*, London: Hodder & Stoughton.
Buist, W. (1987) 'De Nieuwe Hollandse Waterlinie', *Kampioen* ANWB July/Aug. 16–19.
Burtenshaw, D., Bateman, M. and Ashworth, G.J. (1981) *The City in West Europe*, London: Wiley.
Burton, A.M. (1975) *Urban Terrorism*, London: Leo Cooper.
Button, J.W. (1978) *Black Violence: The Political Importance of the 1960's Riots*, Princeton, N.J.: Princeton University Press.
Cable, L.E. (1987) *Conflict of Myths: the Development of American Counter Insurgency Doctrine and the Vietnam War*, New York: New York University Press.
Campbell, D. (1982) *War Plan UK: The Truth about Civil Defence in Britain*, London: Hutchinson.
Carter, H. (1983) *An Introduction to Urban Historical Geography*, London: Edward Arnold.
Clausewitz, C. (1976) *On War*, Princeton: Princeton University Press.
Cohen, S.B. (1973) *Geography and Politics in a World Divided*, New York: Oxford University Press.
—— (1977) *Jerusalem, Bridging the Four Walls: a Geographical Perspective*, New York: Herzl.
Cole, D.H. (1924) *Imperial Military Geography: General Characteristics of the Empire in Relation to its Defence*, London: Sifton Praed.
Collier, B. (1980) *Arms and the Men*, London: Hamish Hamilton.
Connell, J. (1965) 'Writing about soldiers', *Journal of the Royal United Services Institute*, 221.
Contamine, P. (1980) *War in the Middle Ages*, Oxford: Blackwell.
Corney, A. (1983) 'The Portsmouth fortress', *Journal of the Royal Society of Arts*, 131: 578–86.
Cox, J. (1977) *Overkill: the Story of Modern Weapons*, Harmondsworth: Penguin Books.
Creveld, M.V. (1977) *Supplying War: Logistics from Wallenstein to Patton*, Cambridge: Cambridge University Press.
Dale, A. (1982) *Historical Preservation in Foreign Countries*, Paris: ICOMOS.
Davidson, R.N. (1981) *Crime and Environment*, London: Croom Helm.
Davies, G. (1987) 'Potted history', *Marxism Today*, 47.
Deane-Drummond, A. (1975) *Riot Control*, London: Thornton Cox.
Department of the Environment (1988) *Admissions to Historic Properties*, London: HMSO.

Dickinson, R.E. (1961) *The West European City*, London: Routledge & Kegan Paul.
Dilley, R.S. (1986) 'Tourist brochures and tourist images', *Canadian Geographer*, 30: 59–65.
Dobby, A. (1978) *Conservation Planning*, London: Hutchinson.
Douglas, J.D. (1983) 'Strategic planning and nuclear insecurity', *Orbis* 27: 667–94.
Douhet, G. (first edn 1921, English edn 1942) *Command of the Air*, New York: Coward McCann.
Downey, F. (1965) *Louisbourg: Key to a Continent*, Englewood Cliffs, N.J.: Prentice Hall.
Dunnigan, J.F. (1980) 'Berlin '85: the enemy at the gates', *Strategy and Tactics*, 79: 4–14.
Elliott, P. (1982) *The Cross and the Ensign: a Naval History of Malta*, London: Grafton Books.
Ellis, J. (1973) *Armies in Revolution*, London: Croom Helm.
—— (1975) *A Short History of Guerrilla Warfare*, Ian Allen, London.
Ellis, W.S. (1983) 'Beirut: up from the rubble', *National Geographic* 163(2): 262–86.
English Tourist Board (1981) *Planning for Tourism in England*, London: ETB.
Erikson, J. (1975) *The Road to Stalingrad*, London: Weidenfeld & Nicolson.
Evans, D.J. and Herbert, D.T. (1989) *The Geography of Crime* London: Routledge.
Faringdon, H. (1989) *Strategic Geography*, London: Routledge.
Firey, W. (1945) 'Sentiment and symbolism as ecological variables', *American Sociological Review*, 10: 40–8.
Fitzgerald, F. (1973) *Fire in the Lake: The Vietnamese and the Americans in Vietnam*, New York: Vintage Books.
Ford, L.R. (1978) 'Urban preservation and the geography of the city in the USA', *Progress in Geography*, 2: 211–238.
Fuller, J.F.C. (1970) *The Decisive Battles of the Western World 480 BC–1757 AD*, vol. 1, London: Paladin.
Garay, M. (1980) 'Le tourisme culturel en France', *Notes et Etudes Documentaires*, Direction de Documentation Française, Paris.
Garlinski, J. (1985) *Poland in the Second World War*, Basingstoke: Macmillan.
Gellner, J. (1974) *Bayonets in the Streets*, Toronto: Collier Macmillan.
Georges, D.E. (1978) *Geography of Crime and Violence: A Spatial and Ecological Perspective*, Research papers 78.1, Albany: State University of New York.
Gilbert, M. (1976) *The Arab–Israeli Conflict: Its History in Maps*, London: Weidenfeld & Nicolson.
Goodey, B. and Ophir, P. (1982) 'Conservation: the necessary myth', in R. Zetter (ed.) *Conservation of Buildings in Developing Countries*, Working paper 60, Oxford: Dept. of Town Planning. Oxford Polytechnic.
Gordon, D.C. (1980) *Lebanon: The Fragmented Nation*, London: Croom Helm.
Graham, B. (1988) 'The town in the Norman colonisation of the British Isles', in D. Denecke and G. Shaw (eds) *Urban Historical Geography*, pp. 37–52, Cambridge: Cambridge University Press.
Graham, D.O. (1983) *Non-nuclear Defence of Cities*, Cambridge, Mass.: Abt Books.
Grant, M. (1974) *The Army of the Caesars*, London: Weidenfeld & Nicolson.
Green, H. (1973) *Battlefields of Britain and Ireland*, London: Constable.
Greene, J.I. (1943) *Clausewitz*, Washington: Longman Green.
Groen, E. (1988) 'Schuilkelders voor de "happy few"', *Nieuwe Geografenkrant*, 12 June: 24.
Gurr, T.R. (1983) 'Characteristics of political terrorism', in M. Stohl (ed.) *The Politics of Terrorism*, New York: Marcel Dekker.
Haan, T.Z. de and Ashworth, G.J. (1984) *Modelling the Seaside Resort: The Example of Great Yarmouth*, Field Studies series 7, Groningen: GIRUG.

Hackett, J. (1982) *The Untold Story*, London: Sidgwick & Jackson.
Hall, T. (1986) *Planung Europaischer Hauptstadte*, Stockholm: Almquist & Wiksell.
Halliday J. and Cumings, B. (1988) *Korea: The Unknown War*, London: Pantheon.
Hardy, D. (1988) 'Historical geography and heritage studies', *Area*, 20(4): 333–8.
Harries, K.D. (1974) *The Geography of Crime and Justice*, New York: McGraw-Hill.
Harris, T. (1987) 'Government and the specialised military town: the impact of defence policy on urban social structure in the nineteenth century', in M. Bateman and R.C. Riley (eds) *Geography of Defence*, London: Croom Helm.
Harvey, A. (1911) *The Castles and Walled Towns of England*, London: Methuen.
Hedges, A.A.L. (1973) *Yarmouth is an Ancient Town*, Great Yarmouth: Great Yarmouth Borough Council.
Heller, H. (1975) 'Zivilschutz und raumplanung', *Plan* 7/8: 15–18.
Herbert, D.T. (1982) *Geography of Urban Crime*, London: Longman.
Herzog, C. (1982) *The Arab–Israeli Wars*, London: Arms and Armour Press.
Hewison, R. (1987) *The Heritage Industry: Britain in a Climate of Decline*, London: Methuen.
Hewitt, K. (1983) 'Place annihilation: area bombing and the fate of urban places', *Annals of Association of American Geographers*, 73(2): 257–84.
Hofmann, H.H. (1977) 'Zur einfuhrung', in H.H. Hofmann (ed.) *Stadt und Militarische Anlagen*, Forschungs und Sitzungsberichte 114, Akademie fur Raumforschung, Hannover: Herman Schroedel Verlag.
Holmes, R. (1985) *Firing Line*, Harmondsworth: Penguin Books.
Home Office (1981a) *Protection of General Public in War Emergency*, Services Circular 3/81, London: HMSO.
—— (1981b) *Domestic Nuclear Shelters: Technical Guidance*, London: HMSO.
Horne, A. (1977) *A Savage War of Peace: Algeria 1954–1962*, London: Macmillan.
Horne, D. (1984) *The Great Museum: The Reinterpretation of History*, London: Pluto.
Houston, J.M. (1953) *A Social Geography of Europe*, London: Duckworth.
Hughes, Q. (1969) *Fortress: Architectural and Military History in Malta*, London: Lund Humphries.
Huizinga, M.H. (1980) *Maple Leaf Up: De Canadese Opmars in Noord-Nederland*, April 1945, Groningen: Niemeijer.
Humble, R. (1980) *Warfare in the Ancient World*, London: Guild Publishing.
International Revolutionary Solidarity Movement (1980) *Towards a Citizen's Militia, Alternatives to NATO and the Warsaw Pact*, Orkney: Cienfuegos Press.
Irving, D. (1981) *Uprising: One Nation's Nightmare, Hungary 1956*, London: Hodder & Stoughton.
Jarowitz, M. (1969) 'Patterns of collective racial violence', in H.D. Graham and T.R. Gunn (eds) *Violence in America: Historical and Comparative Perspectives*, New York: Bantam.
Jellinek, F. (1971) *The Paris Commune of 1871*, London: Victor Gollancz.
Johnson, C.A. (1962) 'Civilian loyalties and guerrilla conflict', *World Politics*, 14(4): 646–61.
Johnson, P. (1978) *National Trust Book of British Castles*, London: Book Club Associates.
Johnston, A.J.B. (1981) 'Defending Halifax: ordnance 1825–1906', *History and Archaeology* 46, Ottawa: Parks Canada.
Jong, L. de (1970) *Het Koninkrijk der Nederlanden in de Tweede Wereld Oorlong*, vol. 3, *Mei '40*, The Hague: Rijksinstituut voor Oorlogsdocumentatie, Staatsuitgeverij.
Jones, K. (1987) 'Married quarters in England and Wales: a census analysis and

commentary', in M. Bateman and R.C. Riley (eds) *Geography of Defence*, London: Croom Helm.
Jusserand, J.J. (1889) *English Wayfaring Life in the Middle Ages*, London: Methuen.
Kain, R. (1981) *Planning for Conservation: An International Perspective*, London: Mansell.
Kamps, C.T. (1980) 'The central front', *Strategy and Tactics* 82: 4–13.
Karnow, S. (1984) *Vietnam: A History*, Harmondsworth: Penguin Books.
Keegan, J. (1976) *The Face of Battle: A Study of Agincourt, Waterloo and the Somme*, Harmondsworth: Penguin Books.
—— and Holmes R. (1985) *Soldiers: A History of Men in Battle*, London: Hamish Hamilton.
Kennett, V. (1982) *A History of Strategic Bombing*, New York: Scribner.
Kerr, D.G.G. (1960) *An Historical Atlas of Canada*, Toronto: Nelson.
Kightly, C. (1980) *Strongholds of the Realm: Defences in Britain from Prehistory to the Twentieth Century*, London: Thames & Hudson.
King, A.D. (1976) *Colonial Urban Development*, London: Routledge & Kegan Paul.
Kleffens, E.N. (1940) *The Rape of the Netherlands*, London: Hodder & Stoughton.
Koeman-Poel, G.S. (1982) *Bourtange: Schans in het Moeras*, Hoogezand, Holland: Stubey.
Kunzmann, K.R. (1985) 'Military production and regional development in the FRG', *Built Environment*, 11(3): 181–92.
Laqueur, W. (1977a) *Terrorism*, London: Weidenfeld & Nicolson.
—— (1977b) *Guerrilla: A Historical and Critical Study*, London: Weidenfeld & Nicolson.
Lewis, G.M. (1976) 'Geographical aspects of race-related violence in the United States', in J. Wreford Watson and T. O'Riordan (eds) *The American Environment: Perceptions and Policies*, New York.
Lewis, W.J. (1982) *The Warsaw Pact, Arms Doctrine and Strategy*, New York: McGraw-Hill.
L'Huillier, F. (1955) *Histoire de l'Alsace*. Qui sais-je? Paris: Presse Universitaire de France.
Lider, J. (1985) *British Military Thought after World War II*, Aldershot: Gower.
Lindeberg, L. (1974) *De Saedetske Englandskrigene 1801–1814*, Copenhagen: Statsbankerot.
Livingston, M.H. (ed.) (1978) *International Terrorism in the Contemporary World*, Westport, CT: Greenwood Press.
Loewe, M. (1968) *Everyday Life in Early Imperial China*, London: Batsford Books.
Lomax, W. (1976) *Hungary 1956*, London: Allison & Busby.
Lotchin, R.W. (ed.) (1984) *The Martial Metropolis: US Cities in War and Peace*, New York: Praeger.
Lovering, J. (1985) 'Defence expenditure and the regions: the case of Bristol', *Built Environment* 11(3): 193–206.
Lowenthal, D. (1975) 'Past time, present time: landscapes and meaning', *Geographical Review* 65: 1–36.
MacCannell, D. (1976) *The Tourist: A New Theory of the Leisure Class*, New York: Schocken Books.
Mack, A. (1974) 'The non-strategy of urban guerrilla warfare', in J. Niezing (ed.) *Urban Guerrilla War: Studies on the Theory, Strategy and Practice of Political Violence in Modern Societies*, Rotterdam: Rotterdam University Press.
Mackinder, H.J. (1902) *Britain and the British Seas*, Oxford: Clarendon Press.
McNamara, R.S. (1986) *Blundering into Disaster: Surviving the First Century of the Nuclear Age*, New York: Pantheon.

Maguire, T.M. (1899) *Outlines of Military Geography*, Cambridge: Cambridge University Press.
Mahan, A.T. (1890) *The Influence of Sea Power upon History* (republished 1980), London: Bison.
Mao Tse-tung (1963) *Selected Military Writings of Mao Tse-tung*, Peking: Foreign Languages Press.
Markusen, A. (1985) 'The military remapping of the United States', *Built Environment* 11(3): 171–80.
Markusen, A.R. and Bloch, R. (1985) 'Defence cities: military spending, high technology and human settlements', in M. Castells (ed.) *High Technology, Space and Society*, Urban Affairs 28.
Marighella, C. (1971) *Mini-manual of the Urban Guerrilla*, London: International Institute for Strategic Studies.
Mawby, R.I. (1984) 'Vandalism and public perceptions of vandalism in contrasting residential areas', in C. Levy-Leboyer (ed.) *Vandalism*, pp. 235–45, Amsterdam: North-Holland Press.
McCuen, J.J. (1966) *Art of Counter-Revolutionary War*, London: Faber.
Mellor, R.E.H. (1978) *The Two Germanies: A Modern Geography*, London: Heinemann.
—— (1987) 'National defence: the military aspects of political geography', *O'Dell Memorial Monograph* 19, Aberdeen: Department of Geography, University of Aberdeen.
Mercillan, P. (198?) *La Guerre du Liban 1982* Special connaissance de histoire, Paris: Hachette.
Messenger, C. (1976) *The Art of Blitzkrieg*, London: Ian Allen.
Methuin, E.H. (1970) *The Riot Makers*, New Rochelle, N.Y.: Burlington House.
Mickolus, E. (1983) 'Trends in transnational terrorism', in M.H. Livingston (ed.) *International Terrorism in the Contemporary World*, Westport, Conn.: Greenwood Press.
Middlebrook, M. (1980) *The Battle of Hamburg*, Harmondsworth: Penguin Books.
Miranda, J. (1988) 'Revolution in Latin America' *Strategy and Tactics*, 120: 21–59.
Mitchell, J.B. (1955) *Decisive Battles of the Civil War*, New York: Fawcett.
Mitchell, W. (1930) *Skyways*, New York: Lippincott.
Morreau, L.J. (1979) *Bolwerk der Nederlanden: de Vestingwerken van Maastricht*, Assen, Holland: Van Gorcum.
Morris, A.J. (1972) *The History of Urban Form*, London: George Godwin.
Moss, R. (1971) 'Urban guerrilla warfare', *Adelphi Papers* no. 79, London: International Institute for Strategic Studies.
—— (1972) *The War for the Cities*, New York: Coward, McCann & Geoghegan.
Mumford, L. (1961) *The City in History*, Harmonsworth: Penguin Books.
Nader, G.A. (1976) *Cities of Canada*, Toronto: Macmillan.
Nelson, W.A. (1984) *The Dutch Forts of Sri Lanka: the Military Monuments of Ceylon*, Edinburgh: Canongate.
Newman, D. (1972) *Defensible Space*, New York: Macmillan.
Netherlands Research Institute for Tourism (1988) *Tourism Trend Report 1988*, Breda: NRIT.
Ney, V. (1958) 'Guerrilla war and modern strategy', *Orbis* 2(1): 66–82.
Niezing, J. (ed.) (1974) *Urban Guerrilla War: Studies on the Theory, Strategy and Practice of Political Violence in Modern Societies*, Rotterdam: Rotterdam University Press.
Noordegraaf, L. (1985) *Nederlandse Marktsteden*, Amsterdam: Spectrum.
O'Loughlin, J. and van de Wusten, H. (1986) 'Geography, war and peace', in *Progress in Human Geography*, 10(4), London: Arnold.

O'Sullivan, P. and Miller, J.W. (1983) *Geography of Warfare*, London: Croom Helm.
Oatts, L.B. (1949) *Guerrilla Warfare*, Royal United Services Institute, Pub. no. 194: 192–6.
Ommen Kloek, W.K.J.J. van (1947) *De Bevrijding van Groningen*, Assen: Van Gorcum.
Openshaw, S., Steadman, P. and Greene, O. (1983) *Doomsday: Britain After Nuclear Attack*, Oxford: Blackwell.
—— and Steadman, P. (1985) 'Domesday revisited', in A. Pepper and A. Jenins (eds) *Geography of Peace and War*, Oxford: Blackwell.
Peach, C. (1985) 'Immigrants and the 1981 urban riots in Britain', in P.E. White and B. van der Knaap (eds) *Contemporary Studies in Migration*, Norwich: Geobooks.
Pearce, D.G. (1987) *Tourism Today: A Geographical Analysis*, Harlow: Longman.
Peltier, L.C. and Pearcy, G.E. (1966) *Military Geography*, Princeton, N.J.: Van Nostrand.
Penck, A. (1916) 'Der krieg und das studium der geographie', *Zeitschrift der Gesellschaft fur Erdkunde zu Berlin*, 3(4).
Pepper, D. and Jenkins, A. (1985) *Geography of Peace and War*, Oxford: Blackwell.
Philips, C. and Ross, I. (1983) *The Nuclear Casebook: An Illustrated Guide*, Edinburgh: Polygon.
Piers, H. (1947) *The Evolution of the Halifax Fortress*, 1749–1928, Halifax: Public Archives of Nova Scotia.
Platt, R.H. (1984) 'The planner and nuclear crisis relocation', *Journal of the American Planning Association* 50: 259–60.
Pred, A. (1984) 'Place as historically contingent process: structuration and the time geography of becoming places', *Annals of the Association of American Geographers* 74(2): 279–97.
Raisz, E. (1964) *Atlas of Florida*, Gainesville: University of Florida Press.
Ratzel, F. (1903) *Politische Geographie: die Geographie der Staaten, des Verkehres und des Krieges*, Munich/Berlin: Oldenburg Verlag.
Rigg, R.T. (1968) *A Military Appraisal of the Threat to US Cities*, Washington: US Army.
Riley, R.C. (1972) 'The growth of Southsea as a naval satellite and Victorian resort', *Portsmouth Papers* 16, Portsmouth: Portsmouth City Council.
—— (1985) 'The evolution of the docks and industrial buildings in Portsmouth Royal Dockyard 1698–1914', *Portsmouth Papers* 44, Portsmouth: Portsmouth City Council.
—— (1987) 'Military and naval land use as a determinant of urban development: the case of Portsmouth', in M. Bateman and R.C. Riley (eds) *Geography of Defence*, pp. 52–81, London: Croom Helm.
Rolf, R. and Saai, P. (1986) *Vestingwerken in West Europa*, Weesp: Fibula.
Rule, J.B. (1989) *Theories of Civil Violence*, Los Angeles: University of California Press.
RUSIDS (Royal United Services Institute for Defence Studies) (1982) *Nuclear Attack: Civil Defence*, Oxford: Brasseys.
Rustin, R. (1980) 'Tito and his partisan army: Yugoslavia 1941–5', *Strategy and Tactics* 81: 1–12.
Rijksplanologische Dienst (RPD) (1985) *Samenleving en Criminaliteit*, The Hague: Staatsuitgeverij.
Salert, B. and Sprague, J. (1980) *The Dynamics of Riots*, Ann Arbor: Inter-university Consortium for Political and Social Research.
Salch, C.L. (1978) *L'Atlas des Villes et Villages Fortifiés en France*, Strasbourg: Publitotal.
Satoh, S. (1986) 'Innovation in town planning problems and methods in old Japanese

castle towns since the Meiji restoration (1868)', *Proceedings World Planning and Housing Conference*, Adelaide.

Saunders, A.D. (1967) 'Hampshire coastal defence since the introduction of artillery', *Archeological Journal*, 123; 136–50.

Schliemann, H. (1880) *Mycenae*, New York: Arno Press.

Schneider, J. and Patton, W. (1985) 'Urban and regional effects of military spending: a case study of Vallejo, California and Mare Island shipyard', *Built Environment* 11(3): 207–18.

Schuitema Meijer, A.T. (1974) *Groningen in Prent*, Zaltbommel: Europees Bibliotheek.

Selier, H. (1988) 'Een decor in Verval', *NRC Handelsblad* 17/6/88, p. 917.

Sheehan, M. (1983) *The Arms Race*, Oxford: Martin Robertson.

Short, J. (1981) 'Defence spending in the United Kingdom regions', *Regional Studies* 15: 101–10.

Skovmand, S. (1980) *Danmarks Historie fra Landboreformerne til Forsts Verdenskrig*, Copenhagen: Munksgaards.

Sicken, B. (1977) 'Historische entwicklung in stadtraum', in H.H. Hofmann, (ed.) *Stadt en Militarische Anlager*, Forschungs und Sitzungsberichte, 114, Akademie fur Raumforschung, Hannover: Herman Schroedel Verlag.

Silver, C. (1984) 'Norfolk and the navy: the evolution of city federal relations 1917–46', in R.W. Lotchin (ed.) *The Martial Metropolis: US Cities in War and Peace*, pp. 109–34, New York: Praeger.

Slessor, J. (1954) 'Air power and the future of war', *Royal United Services Institute Journal*, August: 11–12.

Smook, R.A.F. (1984) *Binnensteden Veranderen: Atlas van het Ruimtelijk Veranderingsproces van Nederlandse Binnensteden in de Laatste Anderhalve Eeuw*, Zutphen: De Walburg Pers.

Sneep, J.H., Treu, A. and Tijdeman, M. (1982) *Vesting: vier eeuwen vestingbouw in Nederland*, The Hague: Stichting Menno de Coehoorn.

Spaight, J.M. (1930) *Air Power and the Cities*, London: Longman.

Stanley, G.F.G. (1973) *Canada Invaded*, 1775–1776, Ottawa: Canadian War Museum.

Stein, J.M. (1985) 'Militarism as a domestic planning issue', *International Journal of Urban and Regional Research*, 9(3): 341–51.

Stevens, H. (1987) *Hergebruik van Oude Gebouwen*, Heemschut-serie Terra, Baarn: Heemschut.

Stohl, M. (1983) 'Myths and realities of political terrorism', in M. Stohl (ed.) *The Politics of Terrorism*, New York: Marcel Dekker.

Sumner, B.H. (1944) *Survey of Russian History*, London: Methuen.

Sutcliffe, A. (1970) *The Autumn of Central Paris: The Defeat of Town Planning*, London: Edward Arnold.

Tammeling, B. (1980) 'Geschonden stad', *Groningen Toen*, pp. 96–127, Groningen: Groningen Gezinsbode.

Targ, H.R. (1983) 'Societal structure and revolutionary terrorism', in M. Stohl (ed.) *The Politics of Terrorism*, New York: Marcel Dekker.

Taylor, G. (1949) *Urban Geography*, London: Methuen.

Temple-Patterson, A. (1967) '"Palmerston's folly", the Portsdown and Spithead Forts', *Portsmouth Papers* 3, Portsmouth: Portsmouth City Council.

Todd, D. (1980) 'The defence sector in regional development', *Area* 12(2): 115–21.

Tuan, Yi-Fu (1979) *Landscapes of Fear*, Oxford: Blackwell.

Tunbridge, J.E. (1984) 'Whose heritage to conserve? Cross-cultural reflections upon political dominance and urban heritage conservation', *Canadian Geographer* 28(2) 171–80.

Tunbridge, J.E. (1987) 'Conserving the naval heritage: its role in the revitalisation of North American urban waterfronts', in R.E. Riley (ed.) *Urban Conservation: International Contrasts*, Portsmouth Polytechnic, Department of Geography, Occasional Papers 7.

Turnbull, C.M. (1977) *A History of Singapore* 1819–1975, Oxford: Oxford University Press.

Tyrell, I. (1983) *The Survival Option: a Guide to Living Through Nuclear War*, London: Jonathan Cape.

Vance, J.E. (1977) *This Scene of Man*, New York: Harper Row.

Vlerk, L.C. v.d. (1983) *Utrecht Ommuurd*, Vianen: Kwadraat.

Vrankrijker, A.C.J. (1965) *De Historie van de Vesting Naarden*, Naarden: De Haan.

Wanklyn, H. (1961) *Friedrich Ratzel: a Biographical Memoir and Bibliography*, Cambridge: Cambridge University Press.

Weber, M. (1958) *The City*, Illinois: Glencoe.

Wells, P. (1987) 'The military scientific infrastructure and regional development', *Environment and Planning*, A, 19: 1631–58.

Welsenes, C. van (1976) 'De Duits verdediging van de stad Groningen aan het einde van de Tweede Wereld Oorlog', *Groningse Volksalmanak*, Groningen, pp. 92–109.

Wheatcroft, A. (1983) *The World Atlas of Revolution*, London: Hamish Hamilton.

White, P. (1984) *The West European City: A Social Geography*, London: Longman.

Whitehouse, R. (1977) *The First Cities*, Oxford: Phaidon.

Whiting, C. (1987) *The Three Star Blitz: The Baedeker Raids and the Start of Total War*, London: Cooper.

Wilson, J.J.C.P. (1934) *Militaire Aardrijkskunde*, The Hague: Staatsuitgeverij.

Ziegler, D.J. (1985) 'Geography of civil defence', in D. Pepper and A. Jenkins (eds) *Geography of Peace and War*, Oxford: Blackwell.

Zuydewijn, N. de R. van (1977) *Verschanste Schoonheid; een Verassende Ontdekkingstocht Langs Historische Verdedigingswerken in Nederland*, Van Kempen, Amsterdam.

—— (1983) *Neerlands Veste*, The Hague: Staatsuitgeverij.

Index of places

Maps and illustrations are indicated by emboldened page numbers. Names of specific wars are included in the subject index.

Subject index

Maps and illustrations are indicated by emboldened page numbers. Names of places are not included (see index of places) except where they are within the name of a war, e.g. Russo–Japanese.

164276

LIVERPOOL HOPE
UNIVERSITY COLLEGE
LIBRARY
HOPE PARK
LIVERPOOL, L16 9JD